Springer Texts in Education

Springer Texts in Education delivers high-quality instructional content for graduates and advanced graduates in all areas of Education and Educational Research. The textbook series is comprised of self-contained books with a broad and comprehensive coverage that are suitable for class as well as for individual self-study. All texts are authored by established experts in their fields and offer a solid methodological background, accompanied by pedagogical materials to serve students such as practical examples, exercises, case studies etc. Textbooks published in the Springer Texts in Education series are addressed to graduate and advanced graduate students, but also to researchers as important resources for their education, knowledge and teaching. Please contact Yoka Janssen at Yoka.Janssen@springer.com or your regular editorial contact person for queries or to submit your book proposal.

Lisa Bosman • Katey Shirey
Stephanie Fernhaber

Developing the Entrepreneurial Mindset in Engineering Students through a Perspective-Taking Approach

A Toolkit for Curriculum Development, Assessment, and Dissemination

Springer

Lisa Bosman
West Lafayette, IN, USA

Stephanie Fernhaber
Indianapolis, IN, USA

Katey Shirey
Washington, D.C., USA

ISSN 2366-7672 ISSN 2366-7680 (electronic)
Springer Texts in Education
ISBN 978-3-032-15122-3 ISBN 978-3-032-15123-0 (eBook)
https://doi.org/10.1007/978-3-032-15123-0

This Springer imprint is published by the registered company Springer Nature Switzerland AG
The registered company address is: Gewerbestrasse 11, 6330 Cham, Switzerland

Contents

Chapter 1
Introduction

Abstract Engineering education faces a range of persistent challenges, including low student retention and limited transitions into the engineering workforce. This book offers a practical, research-informed response that places engineering faculty at the forefront of addressing these issues. It introduces a perspectives-taking approach to integrating the entrepreneurial mindset into core coursework (drawing on tools from bio-inspired design and the arts) to foster student engagement, expand participation, and strengthen the relevance of engineering education. Grounded in Scholarship of Teaching and Learning (SoTL) principles, defined as the systematic study of teaching and learning processes in order to improve education practices and share findings to advance student learning across disciplines, this book offers both the rationale and resources, including course-based implementation examples, to support faculty-led change in engineering education.

Keywords Engineering faculty · Perspectives-taking approach · Scholarship of teaching and learning (SoTL)

1.1 Why a New Approach Is Needed in Engineering Education

"Times change, we need to change as well."
—*Nelson Mandela*

Engineering education stands at a **pivotal moment**, with increasing recognition that it must better reflect the aspirations, identities, and needs of today's students. While the field has contributed significantly to technological advancement, challenges persist in student retention and workforce alignment. Fewer than 60% of students who begin engineering programs complete their degrees within 6 years (American Society for Engineering Education, 2016), and of those who do, a substantial number do not pursue a related career path (National Science Foundation, 2021). These patterns point to a broader issue: Many students are not finding the

L. Bosman et al., *Developing the Entrepreneurial Mindset in Engineering Students through a Perspective-Taking Approach*, Springer Texts in Education,
https://doi.org/10.1007/978-3-032-15123-0_1

educational experience meaningful, relevant, or supportive enough to sustain their commitment to engineering as a career path (Louten, 2024).

A growing body of research suggests that teaching and learning practices play a central role in this disconnect (Eichler & Gradwohl, 2021). Engineering programs often emphasize rigorous, content-heavy curricula with limited opportunities for creative, collaborative, or socially relevant learning. Faculty, though deeply knowledgeable in their discipline, receive minimal preparation in inclusive or student-centered pedagogy. These challenges are amplified for students from historically underrepresented backgrounds, who often face additional barriers to participation and belonging.

Addressing these challenges requires more than minor adjustments to curricula. Rather, it calls for rethinking the way engineering is taught and whom it serves, and how faculty can shape their content and classrooms for engaging, inclusive, and relevant learning environments.

1.2 Reframing the Role of Engineering Faculty

Faculty are central to the student experience. Through the design of learning environments, faculty influence not only what students learn but also how students engage with and relate to the field of engineering. Yet many engineering educators receive little formal preparation in teaching and often work in systems that prioritize and reward research more than successful classroom teaching.

Empowering faculty to lead change involves more than offering professional development. **It requires reframing their role as agents of transformation in engineering education through instructional shifts and contributing to the Scholarship of Teaching and Learning (SoTL)**. This book supports that shift by introducing a framework and set of tools that help faculty integrate the entrepreneurial mindset through a perspectives-based approach. This includes drawing on fields such as bio-inspired design and the arts to foster relevance, creativity, and broader participation. We provide a step-by-step guide to sharing the outcomes of classroom interventions through faculty reflection, analysis, and publication.

By equipping faculty to design more inclusive and meaningful learning experiences, we position them as leaders in transforming engineering education from the inside out. This transformation extends beyond the individual classroom, as faculty contribute to a broader movement by engaging in SoTL. Through systematic reflection, analysis, and dissemination of their pedagogical innovations, faculty not only improve their own practice but also shape the future of engineering education. In doing so, they become part of a growing community that values teaching as a scholarly endeavor—advancing knowledge, fostering equity, and inspiring change across the field.

1.3 Building on a Foundation

This book is the third in a series focused on advancing how the entrepreneurial mindset is taught in higher education. The first book, *Teaching the Entrepreneurial Mindset to Engineers (Bosman & Fernhaber,* 2018*)*, introduced the entrepreneurial mindset as a critical competency for engineering students. It provided a foundational understanding of why the entrepreneurial mindset matters and offered practical strategies for integrating the mindset into existing engineering courses, along with examples and resources to support early implementation.

The second book, *Teaching the Entrepreneurial Mindset Across the University (Bosman & Fernhaber,* 2021*)*, broadened the conversation beyond engineering. It introduced the Entrepreneurial Mindset Teaching Blueprint—a structured approach for embedding mindset development into existing coursework—and demonstrated how faculty across disciplines could adapt entrepreneurial thinking to their own fields. This integrative approach addressed barriers related to credit limitations, faculty training, and student access, making entrepreneurial learning more inclusive and accessible across campus.

This third book returns to engineering education while introducing a new contribution: the integration of the entrepreneurial mindset with a ***perspectives-taking approach***. Drawing in fields such as bio-inspired design and the arts, this approach encourages students to connect technical learning with creativity, human-centered problem solving, and nature. By expanding how we frame and teach the entrepreneurial mindset, this book offers strategies for broadening participation, deepening relevance, and reimagining the role of engineering education. It also introduces a new set of tools—grounded in SoTL—to help faculty move from idea to implementation, assessment, and dissemination of their teaching innovations. SoTL involves rigorously investigating and reflecting on teaching practices to enhance student learning and openly sharing the results to contribute to the broader educational community.

Figure 1.1 offers an overview of our conceptual framework, which suggests that preparing engineering graduates requires both the integration of the entrepreneurial mindset into the curriculum coupled with perspective taking in order to make the content highly engaging and relevant.

1.4 What to Expect in This Book

This book is designed as both a guide and a resource for engineering educators who are ready to rethink how they teach and reach students. It is organized into four parts, as outlined in Fig 1.2.

Part I explores the systemic challenges facing engineering education, including low retention, limited teaching preparation, and narrow definitions of success. Chapter 2, in particular, delves into the data that expose how these enduring issues

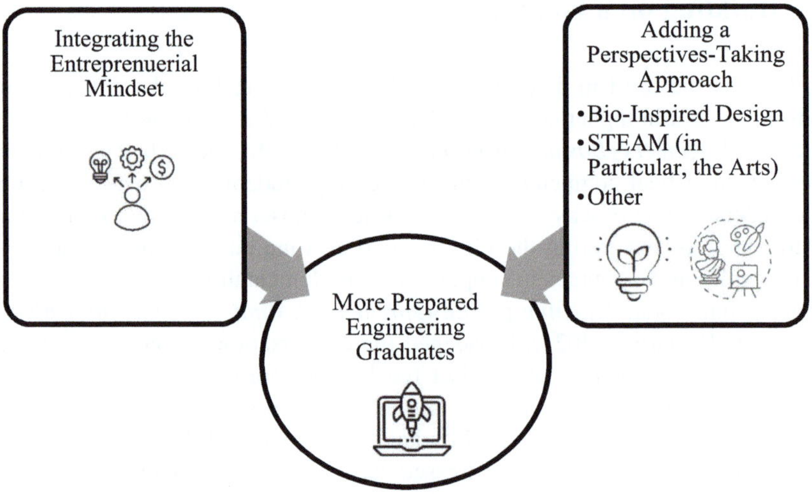

Fig. 1.1 Our perspectives-taking approach on how bio-inspired design and STEAM can complement and better prepare engineering graduates for the workforce

continue to shape and restrict the field's progress. Following the data-driven insights of Chap. 2, Chap. 3 explores the realities of teacher preparedness in engineering education, uncovering the gaps in pedagogical training and support that hinder meaningful learning experiences. Chapter 4 provides an overview of how to apply an integrative approach to developing students' entrepreneurial mindset as a way to improve teaching and learning in engineering, and provides motivation for coupling such integration with bio-inspired design and the arts.

Part II revisits the integrative approach for embedding the entrepreneurial mindset into core coursework—a foundation introduced in our earlier work. Chapter 5 defines the entrepreneurial mindset and explains why it is essential for preparing engineers to thrive in complex, ever-changing professional environments. Expanding on the foundational concept and motivation for teaching the entrepreneurial mindset, Chap. 6 details how the integrative approach operationalizes the entrepreneurial mindset within core engineering coursework, illustrating practical strategies for curriculum design and implementation.

The book's central contribution is introduced in Part III: a perspectives-taking approach that leverages bio-inspired design and the arts to foster inclusion, creativity, and broader participation in engineering. Chapter 7 introduces the perspectives-taking approach, explaining how it broadens students' thinking and enhances their ability to address complex, real-world problems. Expanding on the perspectives-taking approach, Chap. 8 examines bio-inspired design as a powerful framework for fostering creativity, innovation, and systems thinking within engineering education. Chapter 9 highlights the value of incorporating the arts into STEM—demonstrating how STEAM education nurtures empathy, creativity, and the holistic problem-solving essential for the engineers of the future.

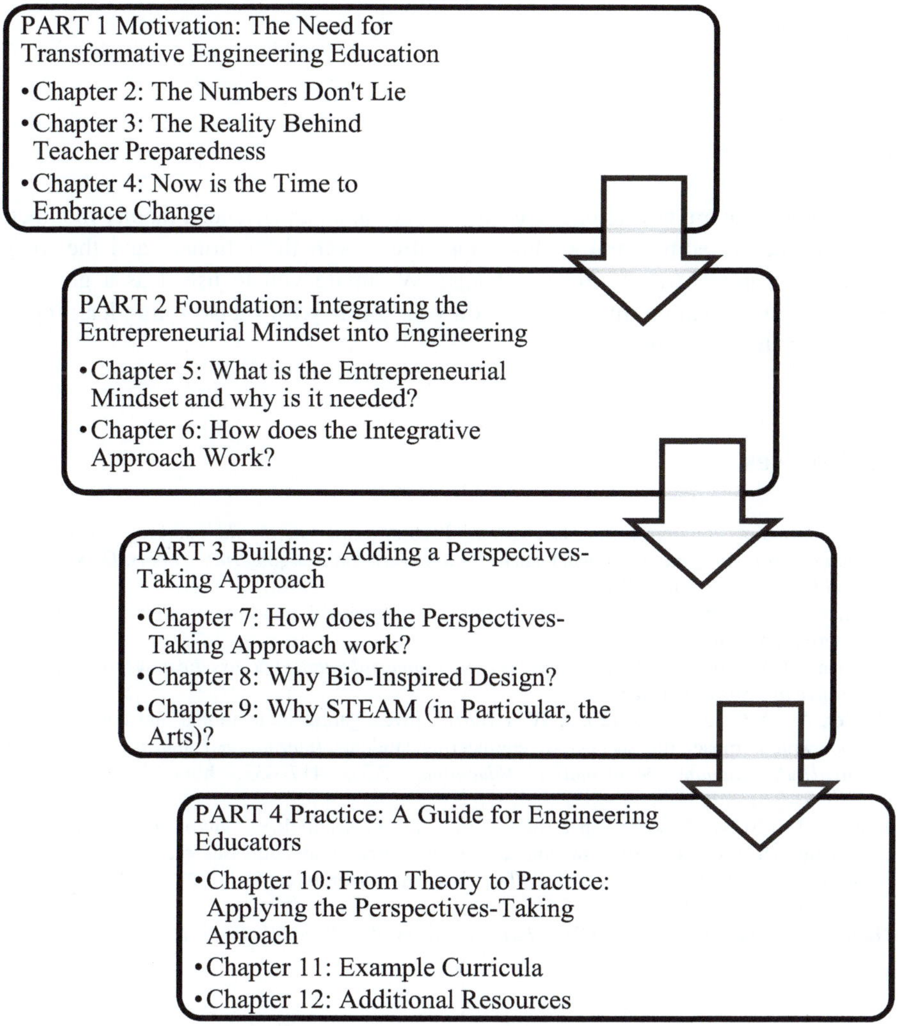

Fig. 1.2 Progression of this book.

Finally, Part IV provides practical guidance for faculty implementation. Chapter 10 translates theory into practice by showing how the perspectives-taking approach can be implemented through the Scholarship of Teaching and Learning (SoTL). Building on these practical strategies, Chap. 11 showcases example curricula that illustrate how the perspectives-taking approach and integrative methods can be applied across different engineering contexts. Chapter 12 offers additional resources to help educators further implement, adapt, and sustain these approaches in their own teaching and curriculum development.

SoTL Accelerator

Whether you are new to engineering education, an experienced faculty member, or an academic administrator, this book offers both the rationale and the tools needed to support transformative change. We invite you to use it as a guide, a resource, and a meaningful way point on your journey for transforming learning—starting in the classroom.

References

American Society for Engineering Education. (2016). *Engineering by the numbers: ASEE retention and time-to-graduation benchmarks for undergraduate engineering schools, Departments and Programs*. Brian L. Yoder.

Bosman & Fernhaber. (2018). *Teaching the entrepreneurial mindset to engineers*. Springer International Publishing.

Bosman & Fernhaber. (2021). *Teaching the entrepreneurial mindset across the university: An integrative approach*. Springer.

Eichler, A., & Gradwohl, J. (2021, 2021/10/01). Investigating motivational and cognitive factors which impact the success of engineering students *International Journal of Research in Undergraduate Mathematics Education, 7*(3), 417–437. https://doi.org/10.1007/s40753-020-00127-4.

Louten, J. (2024). Fostering persistence in science, technology, engineering, and mathematics (STEM): Creating an equitable environment that addresses the needs of undergraduate students. *Journal of College Student Retention, 26*(1), 283–302. https://doi.org/10.1177/15210251211073574

National Science Foundation. (2021). *National Survey of College Graduates (NSCG)*.

Part I
Motivation: The Need for Transformative Engineering Education

Chapter 2
The Numbers Don't Lie

Abstract Despite the high demand for skilled engineers, retention and completion rates in engineering programs remain alarmingly low. Many students leave before graduation due to a combination of academic, institutional, and personal challenges. Why are they leaving engineering education and/or not entering the engineering workforce? The literature suggests teaching and learning play a large role in these extreme attrition rates. Factors such as rigorous coursework, inadequate preparation in foundational subjects, lack of engagement, and limited support systems contribute to high attrition rates. Additionally, underrepresented groups, including women and minorities, face unique barriers that further hinder persistence in engineering fields. This chapter provides an overview of the well-documented poor retention rates and lack of diversity in engineering education programs and the engineering workforce, and concludes by providing motivation to improve teaching practices within engineering.

Keywords Engineering education · Engineering workforce broadening participation · Retention rates · Diversity, equity, and inclusion · Completion rates

2.1 Poor Retention, Completion, and Placement in Engineering Education and the Engineering Workforce

> "Effective engineering educators are key to student success; inspiring teaching not only builds knowledge but also motivates students to persist and graduate."
> —*Wendy C. Newstetter, Learning and Cognitive Scientist*

According to the **2021** Engineering by the Numbers Report: ASEE Retention and Time-to-Graduation Benchmarks for Undergraduate Engineering Schools, Departments and Programs, the overall average retention rate for obtaining an engineering degree within **6** years was **55.9%**. According to the **2021** NSF National

L. Bosman et al., *Developing the Entrepreneurial Mindset in Engineering Students through a Perspective-Taking Approach*, Springer Texts in Education,
https://doi.org/10.1007/978-3-032-15123-0_2

Survey of College Graduates, only **65%** of science and engineering college graduates had an occupation related to their highest degree. Putting this into perspective, if **100** students enrolled in an engineering program, **55.9%** (about **56** students) will complete the degree within **6** years. Of those **56** students, **65%** (about **36** graduates) will enter the engineering workforce. In summary, **only about one-third of students who enroll in engineering programs will complete the degree within 6 years AND enter the engineering workforce**. What about the other two-thirds? Why are they leaving engineering education and/or not entering the engineering workforce? The **literature suggests teaching and learning play a large role in these extreme attrition rates** (Feng et al., 2025).

2.2 Broadening Participation in Engineering Is a Strategic Priority

Significant investments in research and development (R&D) are essential for maintaining the competitiveness of the US economy (Ding et al., 2025). Each year, Congress allocates substantial funding through various government agencies to bolster R&D efforts, with the National Science Foundation (NSF) serving as the largest supporter of university-led initiatives. For fiscal years 2023 to 2025, the NSF requested budgets of $9.5 billion, $11.355 billion, and $10.183 billion, respectively (National Science Foundation, 2022, 2023, 2024). The NSF **emphasizes that diversifying participation in STEM fields** through a "dynamic and highly qualified workforce" (National Science Foundation, 2025) is not only a matter of fairness and equity but also a strategic necessity for fostering innovation, economic growth, and societal well-being. The NSF vision of a nation that leads in science and engineering research and innovation, without barriers to participation, underscores the importance of diversity and inclusion as core values. By providing opportunities for all individuals to contribute to and benefit from STEM, we can **cultivate a more inclusive, dynamic, and prosperous future for the nation**.

2.3 DEI Funding vs. Inadequate Teacher Training

The Society of Women Engineers (SWE) Fast Facts 2025 report notes that, as of 2023, **women earned an average of 29% of all bachelor's degrees in engineering, manufacturing, and construction worldwide**. Brazil reported the highest proportion, with women receiving 34% of such degrees. In comparison, the United States awarded 24% of bachelor's degrees in engineering and engineering technology to women in 2022.

Despite gradual increases in female participation, recent research underscores persistent gender disparities in the recruitment and retention of women

in engineering. Studies highlight that women often face barriers such as limited exposure to engineering role models, implicit bias, and classroom environments that do not foster belonging or confidence (Godwin et al., 2016). Retention, meanwhile, is improved through inclusive teaching practices, supportive peer networks, and visible institutional commitment to diversity, equity, and inclusion (Fouad et al., 2017).

While recent efforts to diversify the engineering workforce have been undercut by attacks on diversity, equity, and inclusion (DEI) funding in education, especially in the United States (The White House, 2025), we posit that this is only part of the problem. Another **core issue lies in inadequate teacher preparation**. Research indicates that **attrition rates in engineering programs are exacerbated by the limited training engineering professors** receive in effective teaching and learning strategies (Eichler & Gradwohl, 2021). Whereas K-12 educators undergo rigorous preparation that emphasizes effective teaching practices and students' social-emotional well-being, university faculty often have a minimal amount of required graduate coursework in their subject area, with little emphasis on pedagogical skills (Bosman, 2024). As a result, newly minted PhD graduates enter teaching roles unprepared to address diverse student needs or navigate sensitive topics. **This lack of preparedness not only undermines student retention and completion rates but also hampers efforts to diversify STEM fields**. By examining successful strategies from K-12 teacher education, including ongoing professional development and inclusive teaching practices, higher education institutions can better equip faculty to create supportive and engaging learning environments, ultimately enhancing student success and fostering a more equitable academic landscape.

2.4 Summary

In summary, **the low retention, completion, and placement rates in engineering education underscore the urgent need for systemic changes within the field.** With only about one-third of engineering students completing their degrees and entering the workforce, it is essential to address the underlying issues contributing to these attrition rates. Although significant investments in R&D, particularly through the NSF, emphasize the importance of diversifying participation in STEM, the real challenge lies in the inadequate preparation of engineering faculty to teach effectively. By prioritizing comprehensive teacher training and incorporating best practices from K-12 education, higher education institutions can create more supportive and inclusive learning environments. Such changes will not only improve student engagement and success but also strengthen the engineering workforce, fostering a more innovative and equitable future for the nation.

References

Bosman, L. (2024). Poor teacher training partly to blame for stalled engineering diversity goals. *The Conversation.* https://theconversation.com/poor-teacher-training-partly-to-blame-for-stalled-engineering-diversity-goals-236292. Retrieved 03/15/2025

Ding, Y., Yin, F., Chin, L., Zhou, K., Taghizadeh-Hesary, F., & Li, Y. (2025). Can government R&D expenditure promote innovation? New evidence from 37 OECD countries. *Technological and Economic Development of Economy, 31*(2), 572–596.

Eichler, A., & Gradwohl, J. (2021, 2021/10/01). Investigating motivational and cognitive factors which impact the success of engineering students. *International Journal of Research in Undergraduate Mathematics Education, 7*(3), 417–437. doi:https://doi.org/10.1007/s40753-020-00127-4.

Feng, L., Close, E. W., Luxford, C. J., Pierson, J. A., Olmstead, A., Shim, J., Koka, V. S., & Galloway, H. C. (2025). Transforming undergraduate STEM education: The learning assistant model and student retention and graduation rates. *Research in Higher Education, 66*(1), 4.

Fouad, N. A., Chang, W.-H., Wan, M., & Singh, R. (2017). Women's reasons for leaving the engineering field. *Frontiers in Psychology, 8*, 254941.

Godwin, A., Potvin, G., Hazari, Z., & Lock, R. (2016). Identity, critical agency, and engineering: An affective model for predicting engineering as a career choice. *Journal of Engineering Education, 105*(2), 312–340. https://doi.org/10.1002/jee.20118

National Science Foundation. (2022). *FY 2023 budget request to congress.* https://www.nsf.gov/about/budget/fy2023, Retrieved 03/15/2025.

National Science Foundation. (2023). *FY 2024 budget request to congress.* https://www.nsf.gov/about/budget/fy2024, Retrieved 03/15/2025.

National Science Foundation. (2024). *FY 2025 budget request to congress.* https://www.nsf.gov/about/budget/fy2025, Retrieved 03/15/2025.

National Science Foundation. (2025). *Broadening participation in STEM: Inspiring talent and creating opportunities in science and engineering across the nation.* https://www.nsf.gov/focus-areas/broadening-participation-stem, Retrieved 03/15/2025.

The White House. (2025). *Ending radical and wasteful government DEI programs and preferencing.* https://www.whitehouse.gov/presidential-actions/2025/01/ending-radical-and-wasteful-government-dei-programs-and-preferencing/, Retrieved 03/15/2025.

Chapter 3
The Reality Behind Teaching Preparedness in Engineering

Abstract Most engineering PhD programs focus heavily on research and technical expertise, often neglecting formal training in pedagogy and instructional methods. As a result, many graduates enter academia unprepared for the teaching responsibilities that come with faculty positions. This gap in training affects their ability to design effective courses, engage diverse learners, and assess student outcomes. While some institutions offer teaching workshops or certificate programs, these opportunities are often optional and insufficient. Addressing this gap will enhance the quality of engineering education and support the professional development of new faculty.

Keywords Engineering faculty · Teaching preparedness · Incentives · IRB · Ethics training

3.1 Overview of Limited Teaching Preparedness for Faculty Entering the Engineering Academic Workforce

> "Teachers need support and training to face the complex challenges of today's classrooms; without it, even the best intentions can fall short."
> —*Linda Darling-Hammond, President of the Learning Policy Institute*

Several factors potentially contribute to lack of pedagogical training for engineering PhD programs. First, most disciplinary engineering PhD programs focus more on technical research and development of the dissertation, with little regard for teaching (Shortlidge & Eddy, 2018). PhD students who are awarded a graduate teaching assistantship instead of a graduate research assistantship are typically expected to grade assignments, teach a lab, and/or offer tutoring sessions (Sohoni et al., 2013). Although these tasks are related to teaching, they do not offer a theoretical or holistic perspective of teaching. **Second**, most disciplinary engineering research does not involve human subjects; thus, most disciplinary engineering PhD graduates and faculty members have a limited understanding of institutional review

L. Bosman et al., *Developing the Entrepreneurial Mindset in Engineering Students through a Perspective-Taking Approach*, Springer Texts in Education,
https://doi.org/10.1007/978-3-032-15123-0_3

boards (IRBs)' role in protecting human subjects (Watts & Brightman, 2017). **Third**, many promotion and tenure (P&T) policies do not prioritize teaching efforts; for those that do, focus is often placed on student satisfaction (e.g., end-of-semester course evaluations) rather than student learning (Lattuca et al., 2014). As a result, the limited teaching preparedness of new college and university engineering educators has the potential to turn students off from engineering (Hewitt & Seymour, 1992), which directly impacts retention, completion, and placement rates (Marra et al., 2012).

3.2 Most Disciplinary Engineering Programs Prioritize Research Training over Teacher Training

Given their primary focus on advancing technical knowledge and innovation, most disciplinary engineering programs place greater emphasis on research training than on teacher preparation (Watson & Lyons, 2011). Faculty members in these programs are typically evaluated based on their research output, grant funding, and publications rather than their teaching effectiveness (Freeman et al., 2014). As a result, **graduate students and early-career engineers receive extensive training in conducting experiments, publishing papers, and securing research funding, but they rarely receive formal instruction in pedagogy or classroom management.** This emphasis on research aligns with the broader goal of engineering schools to drive technological progress and industry partnerships, often at the expense of preparing faculty to be effective educators (Brownell & Tanner, 2012; Garcia & Kittur, 2024). Consequently, **many engineering instructors enter academia with little to no formal training in teaching methodologies**, and instead rely on personal experience or mentorship to develop their instructional skills (Agarwal et al., 2020). While some universities offer optional teaching development programs, these are often insufficient to bridge the gap (Felder et al., 2011).

3.3 Most Disciplinary Engineering Programs Have a Limited Focus on IRB and Ethics Training from an Education Perspective

In most engineering disciplines, research training typically excludes work with human subjects, leading to **limited exposure to Institutional Review Boards (IRBs) among PhD graduates and faculty** (Acuña, 2024). This lack of familiarity extends to educational research, where IRBs play a critical role in protecting students and participants in studies on teaching and learning. As engineering educators increasingly engage in education research—such as assessing pedagogical innovations, student learning outcomes, and classroom interventions—they may struggle to navigate IRB requirements and ethical considerations (Subaveerapandiyan et al., 2024). Without proper training, faculty and graduate students risk designing studies that lack appropriate safeguards for human subjects (Gunsalus et al., 2007; Hunckler & Levine, 2022; Singer & Vinson, 2002; Wright et al., 2008).

3.4 Most Disciplinary Engineering Promotion and Tenure Guidelines Fail to Adequately Incentivize Teaching and Learning

Although P&T policies vary widely across universities, **most policies in engineering academia predominantly emphasize research productivity over teaching effectiveness** (Carter et al., 2021; Wankat & Oreovicz, 2003). When teaching is considered, evaluation methods often rely heavily on student satisfaction metrics, such as end-of-semester course evaluations, rather than direct measures of student learning (Johnson et al., 2013; Villanueva et al., 2017). Moreover, students often have the option to participate in course evaluations, which can result in low participation rates (Giesey et al., 2004). This approach can incentivize faculty to prioritize popularity over pedagogical effectiveness, potentially undermining educational quality (Berezvai et al., 2021). Furthermore, reliance on student evaluations has been criticized for biases and limited correlation with actual learning outcomes (Bianchini et al., 2013; Heffernan, 2022).

3.5 Summary

In summary, **disciplinary engineering programs face significant challenges in adequately preparing faculty for their teaching roles.** The predominant emphasis on research training leaves graduate students with insufficient pedagogical skills, hindering their ability to design effective courses and engage students meaningfully. Additionally, the limited focus on IRBs and ethics training related to educational research further compounds this issue, as new educators may lack the necessary understanding to protect human subjects in their studies. Finally, existing P&T guidelines often fail to prioritize teaching effectiveness, and instead rely on student satisfaction metrics that do not accurately reflect learning outcomes. To enhance the quality of engineering education, it is essential for academic programs to integrate comprehensive teacher training, ethical research practices, and robust evaluation methods that recognize and incentivize effective teaching.

Empowering faculty to lead meaningful change in engineering education goes beyond simply providing professional development opportunities. It involves redefining their role to see themselves as proactive change agents who drive innovation through deliberate shifts in instructional practices. This transformation includes not only adopting new teaching methods but also actively engaging in the Scholarship of Teaching and Learning (SoTL), where they critically examine, research, and share effective strategies to improve student outcomes. By embracing this expanded role, faculty become **catalysts for continuous improvement**, shaping the future of engineering education through both practice and scholarship.

References

Acuña, S. A. (2024, 2024/07/01). A practical research methods course that teaches how to be a successful biomedical engineering graduate student. *Biomedical Engineering Education, 4*(2), 295–304. https://doi.org/10.1007/s43683-024-00135-9.

Agarwal, J., Bucks, G., & Murphy, T. J. (2020, 21–24 Oct. 2020). A literature synthesis of professional development programs providing pedagogical training to STEM graduate students. *2020 IEEE frontiers in education conference (FIE).*

Berezvai, Z., Lukáts, G. D., & Molontay, R. (2021). Can professors buy better evaluation with lenient grading? The effect of grade inflation on student evaluation of teaching. *Assessment and Evaluation in Higher Education, 46*(5), 793–808.

Bianchini, S., Lissoni, F., & Pezzoni, M. (2013, 2013/03/01). Instructor characteristics and students' evaluation of teaching effectiveness: Evidence from an Italian engineering school. *European Journal of Engineering Education, 38*(1), 38–57. https://doi.org/10.1080/03043797.2012.742868.

Brownell, S. E., & Tanner, K. D. (2012, 2012/12/01). Barriers to faculty pedagogical change: Lack of training, time, incentives, and … tensions with professional identity? *CBE—Life Sciences Education, 11*(4), 339–346. https://doi.org/10.1187/cbe.12-09-0163.

Carter, R. G., Mundorff, K., Risien, J., Bouwma-Gearhart, J., Bratsch-Prince, D., Brown, S. A., Campbell, A. L., Hartman, J. C., Hasemann, C. A., & Hollenbeck, P. J. (2021). Innovation, entrepreneurship, promotion, and tenure. *Science, 373*(6561), 1312–1314.

Felder, R., Brent, R., & Prince, M. (2011). Engineering instructional development: Programs, best practices, and recommendations. *Journal of Engineering Education, 100*(1), 89–122. https://doi.org/10.1002/j.2168-9830.2011.tb00005.x

Freeman, S., Eddy, S. L., McDonough, M., Smith, M. K., Okoroafor, N., Jordt, H., & Wenderoth, M. P. (2014). Active learning increases student performance in science, engineering, and mathematics. *Proceedings of the National Academy of Sciences, 111*(23), 8410–8415.

Garcia, O. J., & Kittur, J. (2024, 9–12 Dec. 2024). Factors influencing engineering doctoral students' perceptions on their readiness to teach engineering courses. *2024 IEEE international conference on teaching, assessment and learning for engineering (TALE).*

Giesey, J. J., Chen, Y., & Hoshower, L. B. (2004). Motivation of engineering students to participate in teaching evaluations. *Journal of Engineering Education, 93*(4), 303–312. https://doi.org/10.1002/j.2168-9830.2004.tb00819.x

Gunsalus, C. K., Bruner, E. M., Burbules, N. C., Dash, L., Finkin, M., Goldberg, J. P., Greenough, W. T., Miller, G. A., Pratt, M. G., Iriye, M., & Aronson, D. (2007). The illinois white paper: Improving the system for protecting human subjects: Counteracting IRB "Mission Creep". *Qualitative Inquiry, 13*(5), 617–649. https://doi.org/10.1177/1077800407300785

Heffernan, T. (2022). Sexism, racism, prejudice, and bias: A literature review and synthesis of research surrounding student evaluations of courses and teaching. *Assessment and Evaluation in Higher Education, 47*(1), 144–154.

Hewitt, N. M., & Seymour, E. (1992). A long, discouraging climb. *ASEE Prism, 1*(6), 24–28.

Hunckler, M. D., & Levine, A. D. (2022, 2022-September-20). Navigating ethical challenges in the development and translation of biomaterials research [Perspective]. *Frontiers in Bioengineering and Biotechnology, 10.* https://doi.org/10.3389/fbioe.2022.949280.

Johnson, M. D., Narayanan, A., & Sawaya, W. J. (2013). Effects of course and instructor characteristics on student evaluation of teaching across a college of engineering. *Journal of Engineering Education, 102*(2), 289–318. https://doi.org/10.1002/jee.20013

Lattuca, L. R., Bergom, I., & Knight, D. B. (2014). Professional development, departmental contexts, and use of instructional strategies. *Journal of Engineering Education, 103*(4), 549–572. https://doi.org/10.1002/jee.20055

Marra, R. M., Rodgers, K. A., Shen, D., & Bogue, B. (2012). Leaving engineering: A multiyear single institution study. *Journal of Engineering Education, 101*(1), 6–27. https://doi.org/10.1002/j.2168-9830.2012.tb00039.x

Shortlidge, E. E., & Eddy, S. L. (2018). The trade-off between graduate student research and teaching: A myth? *PLoS One, 13*(6), e0199576.

Singer, J., & Vinson, N. G. (2002). Ethical issues in empirical studies of software engineering. *IEEE Transactions on Software Engineering, 28*(12), 1171–1180. https://doi.org/10.1109/TSE.2002.1158289

Sohoni, S., Cho, Y., & French, D. P. (2013). A survey to capture needs assessment for graduate teaching assistant training. *Advances in Engineering Education, 3*(3), n3.

Subaveerapandiyan, A., Taj, A., Ahmad, N., Verma, M. K., & Ahmed, M. G. (2024). Ethics in focus: PhD students' perspectives on IRBs in India. *Global Knowledge, Memory and Communication*, (ahead-of-print). https://doi.org/10.1108/GKMC-04-2024-0234

Villanueva, K. A., Brown, S. A., Pitterson, N. P., Hurwitz, D. S., & Sitomer, A. (2017). Teaching evaluation practices in engineering programs: Current approaches and usefulness. *International Journal of Engineering Education, 33*(4), 1317–1334.

Wankat, P., & Oreovicz, F. (2003, 2003/01/01). Tenure and teaching. *Journal of Professional Issues in Engineering Education and Practice, 129*(1), 2–5. https://doi.org/10.1061/(ASCE)1052-3928(2003)129:1(2).

Watson, J., & Lyons, J. (2011). Aligning academic preparation of engineering Ph. D. programs with the needs of industry. *International Journal of Engineering Education, 27*(6), 1394.

Watts, R. D., & Brightman, A. O. (2017). Crossing the line: When does the involvement of human subjects in testing of engineering capstone design projects require oversight by an IRB? *2017 ASEE annual conference and exposition.*

Wright, D. E., Titus, S. L., & Cornelison, J. B. (2008, 2008/09/01). Mentoring and research misconduct: An analysis of research mentoring in closed ORI cases. *Science and Engineering Ethics, 14*(3), 323–336. https://doi.org/10.1007/s11948-008-9074-5.

Chapter 4
Now Is the Time to Embrace Change

Abstract Now is the time to embrace transformative change in engineering education by implementing innovative approaches that enhance student learning and engagement. First, adopting an integrative approach allows engineering faculty to cultivate the entrepreneurial mindset among students without necessitating the creation of new courses or additional credits, thereby addressing the constraints of already full curricula. Second, employing a perspectives-taking approach, which emphasizes exploring engineering design from a transdisciplinary viewpoint, can broaden participation in the field by fostering a more inclusive, student-centered focus. Finally, disseminating the findings through the Scholarship of Teaching and Learning (SoTL) will contribute to a deeper understanding of effective educational practices and promote continuous improvement in engineering pedagogy. Collectively, these strategies represent a significant opportunity to reshape engineering education and better prepare students for the evolving demands of the profession. This chapter provides an overview of how to apply an integrative approach to developing students' entrepreneurial mindset as a way to improve teaching and learning in engineering, and provides motivation for coupling such integration with bio-inspired design and the arts.

Keywords Integration · Perspectives-taking · Inclusive learning · Assets-based learning · Dissemination · Scholarship of teaching and learning (SoTL) · Engineering education research

4.1 Introduction

> "The illiterate of the twenty-first century will not be those who cannot read and write, but those who cannot learn, unlearn, and relearn."
> —Alvin Toffler, *Future Shock* (Toffler, 1970)

In today's rapidly evolving technological landscape, the urgency for engineering faculty to embrace change in teaching and learning has never been greater. Advances

L. Bosman et al., *Developing the Entrepreneurial Mindset in Engineering Students through a Perspective-Taking Approach*, Springer Texts in Education,
https://doi.org/10.1007/978-3-032-15123-0_4

in technology, shifting industry demands, and increasing diversity among students require educators to move beyond traditional methods and adopt innovative, evidence-based instructional strategies. The pace of change in engineering fields demands that faculty not only stay current with technical knowledge but also continuously refine how they engage and support students. Delaying this transformation risks widening the gap between academic preparation and professional expectations, ultimately impacting student success and the competitiveness of the engineering workforce. Now is the critical moment for engineering educators to lead this change—by integrating active learning, interdisciplinary approaches, and the scholarship of teaching and learning—to better prepare students for the complex challenges they will face. Embracing change today is essential to fostering more inclusive, effective, and future-ready engineering education programs.

4.2 Integrating the Entrepreneurial Mindset into Engineering Curriculum

To better prepare engineering students for a collaborative and convergent workplace, educators should focus on developing students' entrepreneurial mindset in all required engineering coursework (Desing et al., 2022; Jackson et al., 2023). By **integrating the entrepreneurial mindset across the curriculum**, students from all engineering disciplines will be better prepared to enter the workforce, solve complex social issues, and leverage entrepreneurial thinking in their everyday lives (Hylton et al., 2020; Svensson et al., 2020). Engineering faculty can learn to use an integrative approach to teaching the entrepreneurial mindset to engineering students (Jackson et al., 2022). Applying an integrative approach allows engineering faculty to develop students' entrepreneurial mindset in a way that overcomes traditional access issues related to taking more credits, as previously mentioned. **There is no need to create new courses or add credits to engineering programs**, many of which are already at capacity. Furthermore, developing the entrepreneurial mindset is positioned as a required component (versus optional) of completing the course, which thereby increases access to entrepreneurial education. Last, **resources are now available** to aid engineering faculty who are not formally trained in entrepreneurship or in student-centered teaching practices.

4.3 Applying a Perspectives-Taking Approach

Applying a perspectives-taking approach that **explores engineering design through a transdisciplinary lens** holds significant potential for broadening participation in the field of engineering (Gulbulak et al., 2020). This approach encourages collaboration across diverse disciplines, fostering a rich environment where students can engage with complex, real-world problems from multiple viewpoints (Hepburn, 2022; Lattuca et al., 2017). By maintaining a **student-centered focus, the perspectives-taking approach empowers learners** to connect their interests and backgrounds to engineering concepts, enhancing their motivation and involvement (Hernandez-Aguilar et al., 2020). This **inclusive framework** not only prepares students for the multifaceted challenges of modern engineering but also promotes diversity within the discipline (Wächter, 2012), ultimately contributing to a more innovative and effective engineering workforce.

4.4 Disseminating Findings Through SoTL

Developing effective SoTL practices can help support the adoption of active learning practices, which continues to be a challenge in engineering education (Finelli & Froyd, 2019). Moreover, adopting effective SoTL practices **creates a gateway to improved learning and broadening participation**, as engaging in SoTL requires faculty to think more critically as they adopt and disseminate research-based practices.

Potter and Kustra (2011, p. 4) define SoTL as "the systematic study of teaching and learning, using established or validated criteria of scholarship, to understand how teaching (beliefs, behaviors, attitudes, and values) can maximize learning, and/or develop a more accurate understanding of learning, resulting in products that are publicly shared for critique and use by an appropriate community." SoTL is unique in that it is influenced by interdisciplinary settings and centers on practice-driven institutional, curriculum, and classroom investigations that have a clear goal of change (Hubball & Clarke, 2010). SoTL promotes a **science-based teaching methodology** that is frequently motivated by **answering particular, urgent challenges** as opposed to addressing general teaching and learning problems. Although SoTL is often recognized as one of the most rapidly expanding initiatives in academic development within higher education (Divan et al., 2017), challenges remain. First, Boyer's (1990) model of scholarship, which includes four overlapping areas—discovery, integration, application, and teaching—is conceptually confusing. Second, there is continuing discourse related to the operationalization, definition, and intellectual rigidity of SoTL (Boshier, 2009).

The area of study known as Engineering Education Research (EER) produces information with the goal of defining, enlightening, and enhancing engineering education. It is an interdisciplinary field that investigates a wide range of topics pertinent to engineering education by using research approaches from the social sciences and education. The "how" and "why" of engineering education are explored in EER, as opposed to traditional engineering research, which frequently concentrates on technological developments. EER addresses wider issues about the provision of engineering education and goes beyond the boundaries of specific courses. It examines K-12 and postsecondary education as well as professional development to improve engineering education with academics across disciplines, thereby promoting an all-encompassing perspective on engineering education. EER covers a broad spectrum of subjects, including technology integration and curriculum design, where researchers, administrators, and instructors alike are important stakeholders. Researchers provide information on policies and resources, administrators use data to improve programs, and instructors profit from EER's best practices. Case and Light (2011) highlight new approaches in engineering education research while pointing out the disparities in their representation across international domains. Case studies, grounded theories, ethnographies, action research, phenomenography, discourse analyses, and narrative analyses are some of these approaches. In conclusion, EER improves educational procedures and results, thus bridging the gap between engineering and education and increasing the effectiveness and responsiveness of engineering education to the demands of industry and society. SoTL and EER have distinct scopes and research priorities (as explained in Table 4.1).

SoTL investigates student learning and satisfaction based on innovative teaching interventions with the purpose of sharing best practices and lessons learned from an educator's perspective (Bishop-Clark & Dietz-Uhler, 2012). In contrast, EER seeks to understand broader theoretical questions of how and why, investigating factors beyond student learning and satisfaction while considering formal and informal learning contexts. In general, SoTL is considered a more broad, interdisciplinary approach—applicable across academic disciplines—in which theories are *applied*, whereas EER is a specialized field—focused solely on engineering education—in which theories are *created*.

Table 4.1 Factors differentiating SoTL from EER

Factor	SoTL	EER
Scope and focus	Includes research on teaching and learning methods, curriculum building, and educational improvement in this broad approach to educational research, which may be applied to a variety of academic fields.	Specialized area that focuses on enhancing teaching and learning within the framework of engineering disciplines, with a particular focus on engineering education.
Interdisciplinarity	Often incorporates ideas from other disciplines and is impacted by interdisciplinary environments. Not field-specific. Tackles broad concerns related to teaching and learning.	More specialized and multidisciplinary in an engineering setting. Primarily investigates engineering education concerns using research methodologies from the social sciences and education.
Motivation and questions	Driven by the goal of improving instructional strategies and student learning results. Usually tackles more general educational issues and emphasizes teaching that is reflective and grounded in research.	Tackles issues about how engineering is taught and learned, frequently in response to particular obstacles within engineering disciplines. Motivated by the desire to better engineering education from a systems perspective.
Stakeholders	Comprises instructors and learners from a range of disciplines. Study findings frequently apply broadly to a variety of academic contexts.	Incorporates academics, researchers, and administrators in the engineering domain. Results are intended to improve engineering teaching methods.
Methods and approaches	Includes a broad spectrum of qualitative analyses, case studies, surveys, and other research techniques. Encourages introspection on methods of instruction and learning.	Examines engineering education using research methods mostly from the social sciences and education. Engineering context is the main area of emphasis, although case studies, grounded theories, and ethnographies are among the methods used.

4.5 Summary

As we look to **broaden participation in the STEM** fields in general—and engineering, specifically—the limited scholarly training of engineering disciplinary educators leads to a potential challenge for implementing effective pedagogical approaches that are critical for improving representation among populations historically minoritized in STEM.

Integrating SoTL can play a crucial role in enhancing engineering education by applying research-based strategies to improve student learning outcomes, retention, and overall program effectiveness. SoTL can contribute by driving faculty development with evidence-based teaching practices, connecting theory and practice, enhancing student engagement, allowing for curriculum and assessment innovation, and promoting inclusive and equitable education.

By leveraging SoTL, engineering educators can create more effective, engaging, and inclusive learning environments, ultimately improving student retention and success rates in the field.

References

Bishop-Clark, C., & Dietz-Uhler, B. (2012). *Engaging in the scholarship of teaching and learning: A guide to the process, and how to develop a project from start to finish*. Stylus Publishing, LLC.

Boshier, R. (2009). Why is the scholarship of teaching and learning such a hard sell? *Higher Education Research and Development, 28*(1), 1–15.

Boyer, E. L. (1990). *Scholarship reconsidered: Priorities of the professoriate*. ERIC.

Case, J. M., & Light, G. (2011). Emerging research methodologies in engineering education research. *Journal of Engineering Education, 100*(1), 186–210.

Desing, R. M., Kajfez, R. L., Kecskemety, K. M., & Grzybowski, D. M. (2022). Intersections between entrepreneurial minded learning, identity, and motivation in engineering. *International Journal of Engineering Education, 38*(5), 1389–1407.

Divan, A., Ludwig, L. O., Matthews, K. E., Motley, P. M., & Tomljenovic-Berube, A. M. (2017). Survey of research approaches utilised in the scholarship of teaching and learning publications. *Teaching and Learning Inquiry, 5*(2), 16–29.

Finelli, C. J., & Froyd, J. E. (2019). Improving student learning in undergraduate engineering education by improving teaching and assessment. advances in engineering. *Education*.

Gulbulak, U., Ertas, A., & Cordell, M. (2020). The impact of transdisciplinarity on solving complex engineering problems in an ethnically diverse classroom. *International Journal of Engineering Education, 36*(6), 1976–1987.

Hepburn, L.-A. (2022, 2022/05/04). Transdisciplinary learning in a design collaboration. *The Design Journal, 25*(3), 299–316. doi:https://doi.org/10.1080/14606925.2022.2048986.

Hernandez-Aguilar, C., Dominguez-Pacheco, A., Martínez-Ortiz, E., Ivanov, R., Bonilla, J. L. L., Cruz-Orea, A., & Ordonez-Miranda, J. (2020). Evolution and characteristics of the transdisciplinary perspective in research: A literature review. *Transdisciplinary Journal of Engineering and Science, 11*.

Hubball, H., & Clarke, A. (2010). Diverse methodological approaches and considerations for SoTL in higher education. *Canadian Journal for the Scholarship of Teaching and Learning, 1*(1), 2.

Hylton, B. J., Mikesell, D., Yoder, J.-D., & LeBlanc, H. (2020). Working to instill the entrepreneurial mindset across the curriculum. *Entrepreneurship Education and Pedagogy, 3*(1), 86–106.

Jackson, A., Mawson, C., & Bodnar, C. A. (2022). Faculty motivation for pursuit of entrepreneurial mindset professional development. *Entrepreneurship Education and Pedagogy, 5*(3), 320–346. https://doi.org/10.1177/2515127420988516

Jackson, A., Resnick, S., Hansson, R., Burgess, K., & Bodnar, C. A. (2023). Exploration of the experiences that shape engineering students' entrepreneurial mindset development. *Entrepreneurship Education and Pedagogy, 6*(1), 60–86.

Lattuca, L., Knight, D., Seifert, T., Reason, R., & Liu, Q. (2017). Examining the impact of interdisciplinary programs on student learning. *Innovative Higher Education, 2017*, 337. https://doi.org/10.1007/s10755-017-9393-z

Potter, M. K., & Kustra, E. D. (2011). The relationship between scholarly teaching and SoTL: Models, distinctions, and clarifications. *International Journal for the Scholarship of Teaching and Learning, 5*(1), 23.

Svensson, O. H., Adawi, T., Lundqvist, M., & Williams Middleton, K. (2020). Entrepreneurial engineering pedagogy: Models, tradeoffs and discourses. *European Journal of Engineering Education, 45*(5), 691–710.

Toffler, A. (1970). *Future shock, 1970*. Pan.

Wächter, C. (2012). Interdisciplinary teaching and learning for diverse and sustainable engineering education. In *GIEE 2011: Gender and interdisciplinary education for engineers: Formation interdisciplinaire des Ingénieurs et Problème du Genre* (pp. 47–63). Springer.

Part II
Foundation: Integrating the Entrepreneurial Mindset into Engineering

Chapter 5
What Is the Entrepreneurial Mindset and Why Is It Needed?

Abstract The entrepreneurial mindset—defined as the inclination to discover, evaluate, and actively pursue opportunities—is an essential skill set for all students, regardless of discipline. Far from being limited to start-up ventures, entrepreneurial thinking equips engineers to drive innovation, lead continuous improvement, and contribute meaningfully within organizations and in their own lives. These skills enable engineers to navigate the intersection of technological feasibility, customer desirability, and business viability, resulting in designs that deliver value and impact. Consequently, engineering programs are increasingly embracing the Entrepreneurship for All movement. This chapter introduces the entrepreneurial mindset, highlights its relevance to engineering practice, and explores how its integration into engineering education can prepare students to navigate complexity, drive innovation, and create value in diverse contexts.

Keywords Entrepreneurial mindset · Design thinking · Innovation

5.1 The Case for an Entrepreneurial Mindset

> "An entrepreneurial mindset is the ability to see opportunity where others see obstacles, and to create value from uncertainty."
>
> —*Debbie Haski-Leventhal, Professor of Management at Macquarie Business School*

We live in a world marked by rapid change and increasing uncertainty. In the past few years alone, individuals and organizations have had to adapt to a global pandemic, political polarization, advances in artificial intelligence, and an accelerating shift to digital platforms. These disruptions not only transform how we live and work but also acknowledge the limitations of static knowledge and narrowly defined skill sets.

At the same time, we are facing urgent, complex societal challenges such as climate change, global health inequities, food insecurity, and poverty. The United Nation's Sustainable Development Goals outline many of these "wicked problems,"

L. Bosman et al., *Developing the Entrepreneurial Mindset in Engineering Students through a Perspective-Taking Approach*, Springer Texts in Education,
https://doi.org/10.1007/978-3-032-15123-0_5

which require new ways of thinking that are interdisciplinary and oriented toward innovation.

An entrepreneurial mindset offers a compelling response to both disruption and grand challenges. It encourages adaptability, creativity, and a proactive mindset—qualities essential for navigating uncertainty and shaping meaningful change. In this sense, an entrepreneurial mindset is more than a skill set; it is an essential form of literacy for the twenty-first century.

5.2 Understanding the Entrepreneurial Mindset

The term *entrepreneurship* is often associated with start-ups, small business ownership, or rapid-growth ventures. Yet this narrow understanding obscures its broader meaning. To understand the entrepreneurial mindset, we must first revisit the evolution of what it means to be entrepreneurial.

The word *entrepreneur* originates from the French *entreprendre*, meaning "to undertake." Jean Baptiste Say, writing in the early 1800s, described the entrepreneur as a person who moves resources into higher areas of productivity. This early emphasis on **value creation** remains a defining feature of entrepreneurship today. Joseph Schumpeter (1942) later built on this, highlighting the entrepreneur's role in "creative destruction"—bringing about economic progress by displacing the old with the new.

Contemporary definitions have shifted away from context-specific markers, such as business formation, and toward **a process orientation**. Shane and Venkataraman (2000) define entrepreneurship as "the discovery, evaluation, and exploitation of opportunities"—a definition now widely adopted across the field. Peter Drucker (1958) similarly argued that entrepreneurship is a purposeful act of innovation, rooted in the disciplined pursuit of change as opportunity. For Drucker, entrepreneurs do not simply react to change; they actively search for it, respond to it, and exploit it as a source of innovation and impact.

It is this inclination—the capacity to identify, assess, and pursue opportunities—that forms the basis of the entrepreneurial mindset (Bosman & Fernhaber, 2018). It is not limited to those who start companies or take on financial risk. Rather, it represents a **disposition toward action**, a mental framework for navigating complexity, creating value, and adapting to change. De Carolis and Litzky (2019) describe the entrepreneurial mindset as "an innovative approach to thinking and doing" that includes both mindset and skill set and employs initiative, creativity, and calculated risk. Kuratko et al. (2021) further propose three distinct aspects of the entrepreneurial mindset: cognitive, behavioral, and emotional. In recent years, the scholarly conversation on the entrepreneurial mindset has continued to rise (Crosina et al., 2024; Daspit et al., 2023; Shrivastav & Rao, 2025), another indicator of the increased relevancy for educators.

The entrepreneurial mindset can be developed and expressed in a wide range of settings. Students might apply it to launching a new business, but they might also use it to improve processes within an organization, lead change in their community, address global challenges such as climate resilience or public health, or make courageous personal decisions amid uncertainty. This flexibility underscores why the mindset is not confined to any one discipline, career path, or personality type.

Importantly, the entrepreneurial mindset is **learnable**. As cognitive psychologists have long emphasized, our mindsets serve as interpretive filters—shaping what we notice, how we assess situations, and what actions we take (Gupta & Govindarajan, 2002). These filters are not fixed; they evolve through experience, feedback, and reflection. With repeated application—through experimentation, empathy, and value-focused problem-solving—the entrepreneurial mindset becomes a habit of thought. In this way, the entrepreneurial mindset is related to the growth mindset (Dweck, 2006)—in that it can be intentionally developed.

In short, cultivating the entrepreneurial mindset means helping individuals become more **opportunity aware, action oriented, and value driven** in a world that demands adaptability, creativity, and resourcefulness.

5.3 The Engineering Imperative

If an entrepreneurial mindset is essential for navigating complexity and creating value across domains, it is especially vital for engineers. Engineering has always been about designing solutions to real-world problems. Yet in today's rapidly changing world, technical skills alone are no longer enough. The nature of the challenges engineers face—from building resilient infrastructure to developing sustainable technologies—demands more than precision and feasibility. It calls for adaptability, initiative, and value-driven thinking.

This is where the **entrepreneurial mindset** becomes indispensable. Defined as the *inclination to discover, evaluate, and act on opportunities* (Bosman & Fernhaber, 2018), the entrepreneurial mindset complements the engineering skill set by preparing individuals to respond creatively and constructively to ambiguity. While the engineering design process emphasizes problem identification and technical **feasibility**, an entrepreneurial mindset adds layers of **desirability** and **viability (Brown,** 2009). It helps engineers consider not only "Can we build it?" but also "Should we build it?" "Who is it for?" and "How might we deliver it meaningfully and sustainably?"

Engineering roles are evolving. Today's engineers are expected to not only solve technical problems but also contribute to innovation, strategy, and organizational impact. Whether working in start-ups, multinational firms, government research labs, or nonprofit organizations, they must engage with customers, align their work with broader goals, and deliver solutions that generate both economic and social value. An entrepreneurial mindset equips engineers to meet these expectations by helping them identify opportunities, navigate uncertainty, and lead innovation from within complex systems (Weilerstein & Byers, 2016).

Industry expectations have shifted accordingly. Although technical competence remains foundational, employers consistently rank communication, interpersonal skills, problem-solving, and commitment as top priorities—skills strongly associated with an entrepreneurial mindset (Brown, 2009). Engineers who can navigate both technical and contextual complexity add strategic value in any organization.

Perhaps most importantly, an entrepreneurial mindset empowers engineers to address some of society's most pressing and dynamic problems. Climate change, health care inequity, and digital transformation require more than disciplinary expertise. They require **opportunity-oriented thinkers** who can integrate knowledge across boundaries, work in uncertain conditions, and lead change. These are the conditions in which engineers now operate—and where the entrepreneurial mindset becomes not just relevant, but necessary.

In short, engineers are no longer just problem-solvers—they are **opportunity makers**. The entrepreneurial mindset enables them to extend the impact of their technical training, innovate with purpose, and contribute meaningfully to a world that requires both rigor and creativity.

5.4 Transforming Education: Embedding the Mindset

As the entrepreneurial mindset becomes increasingly recognized as essential for navigating today's complex, dynamic world, colleges and universities are reimagining how they prepare students for the future. Once confined to business schools, entrepreneurship is now being embedded across disciplines, from the arts and humanities to public health, education, and the sciences. This shift reflects a broader understanding: Entrepreneurship is not just a career path but a mindset that adds value across sectors.

Engineering education has been central to this transformation. Given the applied nature of the field and its focus on innovation, engineers are uniquely positioned to benefit from—and contribute to—an entrepreneurial mindset. Embedding the mindset into engineering education equips students not only with technical expertise but also with the ability to identify opportunities, work across disciplines, and create solutions that are both innovative and impactful.

5.5 Summary

This chapter examined the growing importance of the entrepreneurial mindset in a world shaped by uncertainty, rapid change, and global challenges. The entrepreneurial mindset was presented as a valuable orientation that extends far beyond business creation, offering relevance in organizational, societal, and personal contexts.

We highlighted how this mindset is particularly vital for engineers, whose contributions increasingly require innovation, adaptability, and systems thinking. As entrepreneurship continues to expand across disciplines, engineering stands out as a field where embedding this mindset can significantly enhance students' capacity to create meaningful impact.

In the next chapter, we explore how this integration is taking shape and what it looks like in practice.

References

Bosman, & Fernhaber. (2018). *Teaching the entrepreneurial mindset to engineers*. Springer International Publishing.

Brown, T. (2009). *Change by design: How design thinking transforms organizations and inspires innovation*. Harper Collins.

Crosina, E., Frey, E., Corbett, A., & Greenberg, D. (2024). From negative emotions to entrepreneurial mindset: A model of learning through experiential entrepreneurship education. *Academy of Management Learning and Education, 23*(1), 88–127. https://doi.org/10.5465/amle.2022.0260

Daspit, J. J., Fox, C. J., & Findley, S. K. (2023, 2023/01/02). Entrepreneurial mindset: An integrated definition, a review of current insights, and directions for future research. *Journal of Small Business Management, 61*(1), 12-44. https://doi.org/10.1080/00472778.2021.1907583.

De Carolis, D. M., & Litzky, B. E. (2019). Unleashing the potential of university entrepreneurship education: A mandate for a broader perspective. *New England Journal of Entrepreneurship, 22*(1), 58–66.

Dweck, C. (2006). *Mindset: The new psychology of success*. Random House.

Gupta, A. K., & Govindarajan, V. (2002). Cultivating a global mindset. *Academy of Management Perspectives, 16*(1), 116–126.

Kuratko, D. F., Fisher, G., & Audretsch, D. B. (2021, 2021/12/01). Unraveling the entrepreneurial mindset. *Small Business Economics, 57*(4), 1681–1691. https://doi.org/10.1007/s11187-020-00372-6.

Schumpeter, J. A. (1942). *Capital. Socialism, and democracy*. Harper.

Shane, S., & Venkataraman, S. (2000). The promise of entrepreneurship as a field of research. *Academy of Management Review, 25*(1), 217–226.

Shrivastav, S., & Rao, P. V. M. (2025, 2025/09/03). Defining, cultivating, and assessing entrepreneurial mindset in higher education: A systematic review and future research directions. *European Journal of Engineering Education, 50*(5), 1126–1159. https://doi.org/10.1080/03043797.2025.2490551.

Weilerstein, P., & Byers, T. (2016). Guest editorial: Entrepreneurship and innovation in engineering education. *Advances in Engineering Education, 5*(1), 1–9.

Chapter 6
How Does the Integrative Approach Work?

Abstract Expanding on the foundational concept and motivation for teaching the entrepreneurial mindset, this chapter details how the integrative approach operationalizes the entrepreneurial mindset within core engineering coursework, illustrating practical strategies for curriculum design and implementation. Despite the relevancy of the entrepreneurial mindset and entrepreneurship training for all career paths, only a small percentage of the higher education student population takes part in entrepreneurially minded learning opportunities. This gap can be attributed to several factors. From a program perspective, many engineering degrees are already at credit capacity which allows limited room in the existing curriculum to add more. From a student's perspective, entrepreneurship education is positioned as optional and requires extra time (and, in some cases, tuition) to do so. From an educator's perspective, engineering faculty are typically not trained in entrepreneurship and may not know where to start. This chapter provides an overview of the integrative approach and how making changes within existing courses can result in greater impact than developing new courses and degree programs.

Keywords Integrative approach · Engineering curriculum · Entrepreneurial mindset

6.1 The Participation Gap in Entrepreneurship Education

> "While the resulting rapid, wide-scale adoption of entrepreneurship programming signals the growing acceptance of entrepreneurship education as a means of cultivating entrepreneurial engineering students, [they] are typically non-compulsory, elective courses, or activities."
>
> —*Shekhar and Huang-Saad* (2021, p. 2)

The entrepreneurial mindset is increasingly vital in engineering, where graduates must navigate ambiguity, anticipate future needs, and create value across diverse contexts. Engineers are no longer expected to solve only technical problems; they

L. Bosman et al., *Developing the Entrepreneurial Mindset in Engineering Students through a Perspective-Taking Approach*, Springer Texts in Education,
https://doi.org/10.1007/978-3-032-15123-0_6

are also called upon to design solutions that are feasible, viable, and desirable (Brown, 2009). Yet despite its relevance, entrepreneurial thinking is absent from most students' formal engineering education. This limited engagement reinforces the marginalization of entrepreneurship within engineering education and limits students' exposure to ways of thinking that are increasingly essential to engineering practice.

A major contributor to this gap is structural. Engineering curricula are often tightly structured, with most credit hours devoted to core technical content, leaving little room to incorporate interdisciplinary material without displacing existing courses (Borrego et al., 2010). Even when an entrepreneurship course is included, a one-time introduction is rarely sufficient to cultivate a new mindset, which develops through ongoing practice and reinforcement over time.

In response to these constraints, many institutions offer optional minors or cocurricular programs in entrepreneurship. Although valuable, these models require additional time or tuition and may inadvertently signal that that entrepreneurial thinking lies outside the core engineering identity (Duval-Couetil et al., 2016). Moreover, because participation is voluntary, these models often attract students who already identify as entrepreneurial, rather than those who might benefit most from such development (Liñán et al., 2018).

6.2 Rethinking the Solution: Integration over Addition

In light of the structural and perceptual barriers, **embedding entrepreneurial thinking within required engineering courses** offers a more inclusive and scalable path forward. This integrated approach reframes entrepreneurship not as a separate subject or career track but as a way of thinking that is aligned with core engineering competencies. By drawing connections between entrepreneurial tools and disciplinary content, the approach allows mindset development to occur within the boundaries of existing curricula. This could involve, for example, incorporating online discussion prompts (Bosman et al., 2019), value creation projects (DeWaters & Kotla, 2023), or leveraging other entrepreneurially minded tools.

Unlike additive models that rely on separate courses or cocurricular programs, integration meets students where they already are—within technical classrooms and project-based environments. It normalizes entrepreneurial thinking as part of effective engineering practice, reaching students who might not otherwise self-select into entrepreneurship-focused offerings.

Integration also addresses a persistent instructional barrier: faculty expertise. Most engineering educators have limited training in entrepreneurship and often feel unprepared to teach it (Shartrand et al., 2008). The integrated approach lowers this threshold by providing a flexible pedagogical framework. Rather than asking faculty to become entrepreneurship experts, it equips them to embed topics such as opportunity identification, value creation, and iterative problem-solving within the scope of their existing teaching. This alignment builds faculty confidence while expanding students' exposure to the critical ways of thinking increasingly demanded in engineering practice.

Crucially, integration is not about adding discrete activities but about shaping how students experience technical content through an entrepreneurial mindset lens. The next section introduces four instructional priorities designed to support faculty in embedding entrepreneurial thinking authentically and sustainably within engineering courses.

6.3 Four Instructional Priorities for Integration

While the integrated approach provides a compelling alternative to traditional entrepreneurship education models, successful implementation depends on how entrepreneurial thinking is introduced and reinforced within technical engineering coursework. To support faculty in this process, Bosman and Fernhaber (2018, 2021) identified four instructional priorities as particularly effective: **entrepreneurial integration, professional skill development, mindset cultivation, and teaching with intention.** These priorities offer adaptable entry points for instructors—regardless of their prior entrepreneurship training—to embed mindset development in ways that enhance, rather than disrupt, existing course objectives.

Entrepreneurial integration. To integrate the *entrepreneurial* aspect of the entrepreneurial mindset into an existing engineering course, there are three core entrepreneurial components that can be drawn upon. First, given that entrepreneurship involves the "processes of discovery, evaluation, and exploitation of opportunities" (Shane & Venkataraman, 2000, p. 218), engineering educators can integrate any part of the entrepreneurial process. This might entail, for example, asking students to analyze the entrepreneurial process, creating exercises where students can directly experience aspects of the process (e.g., searching for and identifying an entrepreneurial opportunity), or assessing one's role in the entrepreneurial process (Fig. 6.1).

Second, to act upon an entrepreneurial opportunity, it must have value. As noted by Brown (2009), the most valuable design exists at the intersection of customer desirability, business viability, and technological feasibility (Fig. 6.2). Whereas engineers traditionally focus heavily on the feasibility aspect, this framework challenges engineering educators to better prepare students to think bigger. Potential integration might include the business model canvas (Osterwalder & Pigneur, 2010), a tool that enables users to visualize an opportunity by considering all these factors.

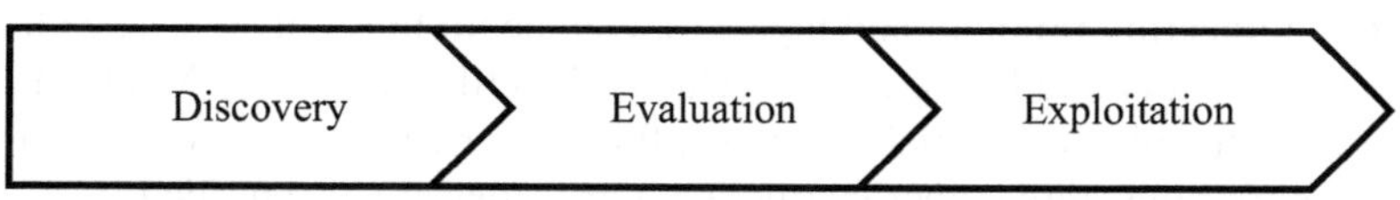

Fig. 6.1 The entrepreneurial process

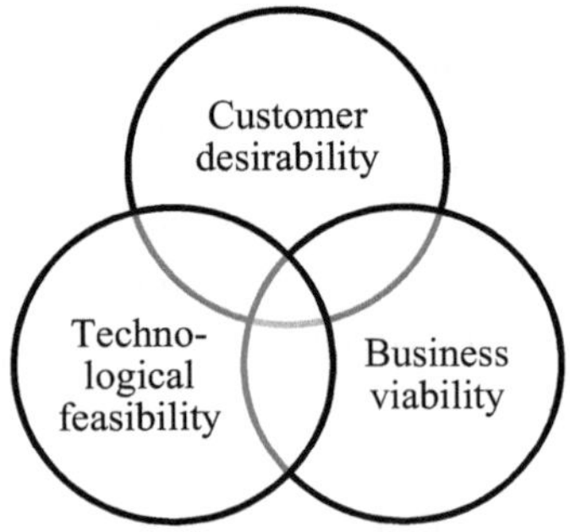

Fig. 6.2 The most valuable design

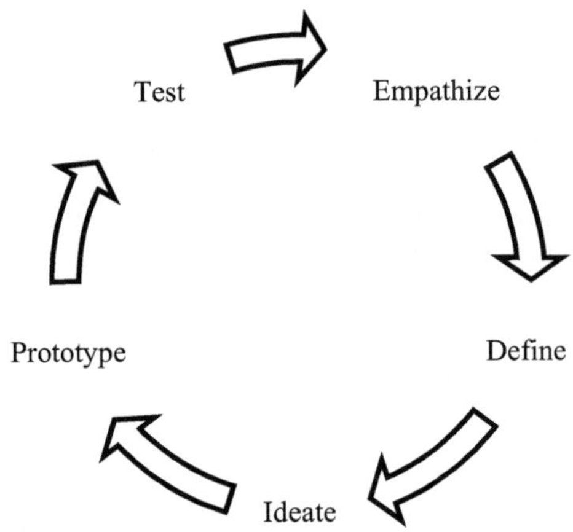

Fig. 6.3 Creative problem-solving

Another option is to conduct market analyses alongside the development of a new technology.

Third, given that opportunities typically involve people, a reliance on creative problem-solving techniques that are people-centered is vital. Design thinking offers an ideal starting point. As put forth by the Stanford d.school, the five-step iterative process (Fig. 6.3) starts by gathering empathy, and then moving on to (re)define the problem, ideate, prototype, and test. Integration into the classroom can involve a deep dive into any one of these problem-solving components—such as the development of an empathy map, creating a low-fidelity storyboard prototype, or using a brainwriting exercise—or involving students in the entire process through a class application project.

When entrepreneurship is integrated in this way—not as a separate track but as a mode of engineering problem-solving—students begin to see its relevance to their disciplinary identity and future careers (Duval-Couetil et al., 2012). While integrating entrepreneurship is essential, it is not sufficient for cultivating the entrepreneurial mindset. For this, three other instructional priorities are also needed.

Professional skill development. Effective engineering solutions rarely emerge in isolation. *Collaboration* and *communication* are consistently ranked among the most critical twenty-first-century skills for engineering graduates (Rios et al., 2020), yet these are often underdeveloped in technically focused curricula. As noted by Evans (2012), collaborative problem-solving enhances both technical innovation and students' capacity to navigate complex team environments. Integrating structured opportunities for teamwork, presentations, and writing into engineering courses reinforces entrepreneurial competencies while giving them "soft skills" to apply them outside of the classroom.

Mindset cultivation. The entrepreneurial mindset is not a fixed trait but a cognitive and behavioral orientation that can be cultivated through repeated practice, feedback, and reflection (Dweck, 2006; Zimmerman & Schunk, 2011). Engineering educators can foster this mindset, for instance, by incorporating scaffolded assignments that encourage experimentation and reward iteration. Feedback is important, whether provided by the instructor, peers, or even students themselves, as it helps students refine their thinking and deepen their learning over time (Winstone & Boud, 2020). Although reflection is typically done on an individual basis, collaborative reflection can also be helpful.

Teaching with intention. Intentional curriculum design ensures that entrepreneurial mindset goals are embedded not only in course activities but also in learning outcomes and assessments. Backward design—starting with desired learning goals and working backward to align instructional methods—has been shown to improve impact in engineering education (Ambrose et al., 2010; Wiggins & McTighe, 2005). Additionally, supporting student motivation through increased self-efficacy, perceived value, and a supportive environment has been empirically linked to improved learning outcomes (Schunk et al., 2014). Instructors who use these strategies position students not just to learn engineering but to apply it entrepreneurially.

Together, these four instructional priorities form the foundation for integrating entrepreneurial thinking in engineering education (Fig. 6.4). They enable faculty to move beyond surface-level exposure and foster deeper, transferable capabilities that prepare students for the uncertain, opportunity-rich environments they will face in their careers.

6.4 Summary

Entrepreneurial thinking is increasingly essential for engineers navigating complex, uncertain, and opportunity-rich environments. However, traditional additive approaches—such as offering electives or cocurricular programs—have not substantially broadened participation or normalized entrepreneurial competencies within core engineering education.

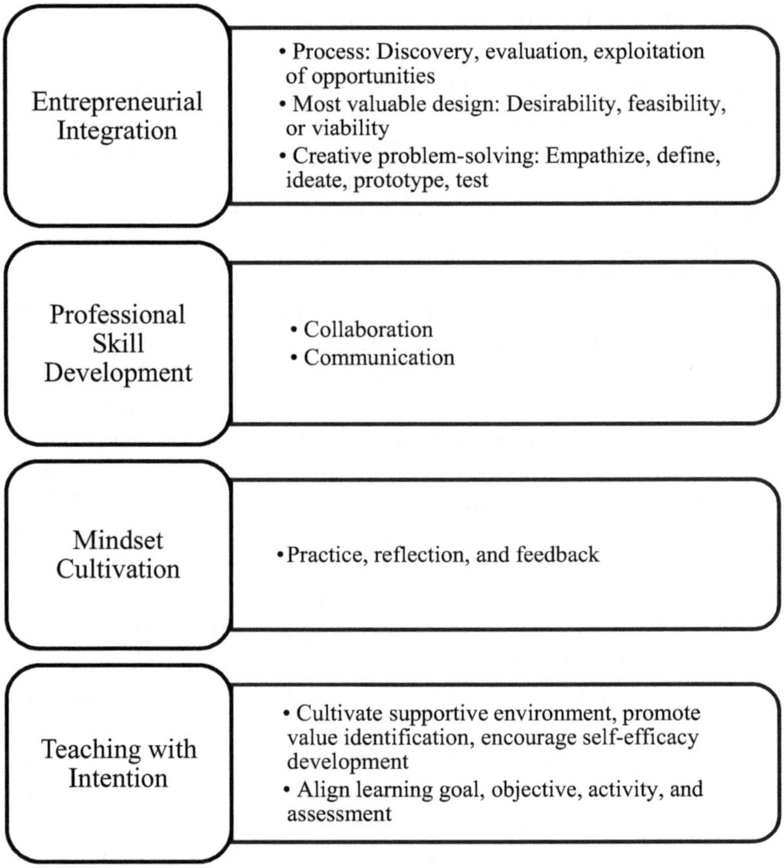

Fig. 6.4 Four instructional priorities for integration

This chapter presented integration as a more scalable and inclusive alternative. By embedding entrepreneurial thinking directly into required courses, faculty can reach a wider range of students and reinforce the idea that entrepreneurship is not a separate track but a mindset relevant to technical problem-solving and professional identity. The integrated approach also supports faculty—many of whom lack formal training in entrepreneurship—by offering pedagogical strategies that align with existing course structures.

To guide implementation, four instructional priorities were introduced: entrepreneurial integration, professional skill development, mindset cultivation, and teaching with intention. Through intentional integration, educators can prepare engineers who are not only technically skilled but also opportunity aware, reflective, and equipped for the real world.

References

Ambrose, S., Bridges, M., Lovett, M., DiPietro, M., & Norman, M. (2010). *How learning works.* Jossey-Bass.

Borrego, M., Froyd, J. E., & Hall, T. S. (2010). Diffusion of engineering education innovations: A survey of awareness and adoption rates in U.S. Engineering Departments. *Journal of Engineering Education, 99*(3), 185–207. https://doi.org/10.1002/j.2168-9830.2010.tb01056.x

Bosman, & Fernhaber. (2018). *Teaching the entrepreneurial mindset to engineers.* Springer International Publishing.

Bosman, & Fernhaber. (2021). *Teaching the entrepreneurial mindset across the university: An integrative approach.* Springer.

Bosman, L., Duval-Couetil, N., Mayer, B., & McNamara, P. (2019). Using online discussions to develop the entrepreneurial mindset in environmental engineering undergraduates: A case study. *International Journal of Engineering Pedagogy.*

Brown, T. (2009). *Change by design: how design thinking transforms organizations and inspires innovation.* Harper Collins.

DeWaters, J., & Kotla, B. (2023, 2023-May-10). Using an open-ended socio-technical design challenge for entrepreneurship education in a first-year engineering course [Curriculum, Instruction, and Pedagogy] *Frontiers in* Education, *Volume 8 - 2023.* https://doi.org/10.3389/feduc.2023.1198161.

Duval-Couetil, N., Reed-Rhoads, T., & Haghighi, S. (2012). Engineering students and entrepreneurship education: Involvement, attitudes and outcomes. *International Journal of Engineering Education, 28*(2), 425.

Duval-Couetil, N., Shartrand, A., & Reed, T. (2016). The role of entrepreneurship program models and experiential activities on engineering student outcomes. *Advances in Engineering Education, 5*(1), n1.

Dweck, C. (2006). *Mindset: The new psychology of success.* Random House.

Evans, N. (2012). Destroying collaboration and knowledge sharing in the workplace: A reverse brainstorming approach. *Knowledge Management Research and Practice, 10*(2), 175–187.

Liñán, F., Ceresia, F., & Bernal, A. (2018). Who intends to enroll in entrepreneurship education? Entrepreneurial self-identity as a precursor. *Entrepreneurship Education and Pedagogy, 1*(3), 222–242.

Osterwalder, A., & Pigneur, Y. (2010). *Business model generation: A handbook for visionaries, game changers, and challengers.* Wiley.

Rios, J. A., Ling, G., Pugh, R., Becker, D., & Bacall, A. (2020). Identifying critical 21st-century skills for workplace success: A content analysis of job advertisements. *Educational Researcher, 49*(2), 80–89.

Schunk, D. H., Meece, J. R., & Pintrich, P. R. (2014). *Motivation in education: Theory, research, and applications* (4th ed.). Pearson Higher Education.

Shane, S., & Venkataraman, S. (2000). The promise of entrepreneurship as a field of research. *Academy of Management Review, 25*(1), 217–226.

Shartrand, A., Weilerstein, P., Besterfield-Sacre, M., & Olds, B. M. (2008, 22–25 Oct. 2008). Assessing student learning in technology entrepreneurship. *2008 38th annual frontiers in education conference.*

Shekhar, P., & Huang-Saad, A. (2021, 2021/06/14). Examining engineering students' participation in entrepreneurship education programs: Implications for practice *International Journal of STEM Education, 8*(1), 40. doi:https://doi.org/10.1186/s40594-021-00298-9.

Wiggins, G. P., & McTighe, J. (2005). *Understanding by design.* ASCD Publications.

Winstone, N. E., & Boud, D. (2020). The need to disentangle assessment and feedback in higher education. *Studies in Higher Education*, 1–12.

Zimmerman, B. J., & Schunk, D. H. (2011). Self-regulated learning and performance. *Handbook of self-regulation of learning and performance* (pp. 1–12).

Part III
Building: Adding a Perspectives-Taking Approach

Chapter 7
How Does the Perspectives-Taking Approach Work?

Abstract Many real-world problems and opportunities require interdisciplinary approaches (e.g., taking different perspectives) for effective solutions. Perspectives-taking helps engineering students see connections between disciplines and understand how technical and nontechnical disciplinary knowledge can be integrated to tackle multifaceted issues. This chapter provides an overview of how valuing multiple perspectives (specifically, bio-inspired design and the arts) alongside the entrepreneurial mindset can support novice engineers and encourage broader participation and retention among persons traditionally underrepresented in engineering, including women and minorities, perhaps due to their being more service oriented and interested in improving the human condition.

Keywords Perspectives · Empathy · Motivation · Engineering · Retention

7.1 The Value of Perspectives-Taking for Novice Engineers

> "By considering multiple problem perspectives varying in scope and focus, designers position themselves to increase their understandings of the 'real' problem and engage in more diverse idea generation processes leading to an increasing variety of potential solutions."
> —*Murray et al.* (2019, p. 248)

Engineering may seem like a black-and-white field, even to engineering students. However, **considering engineering challenges from more than just formulaic, technical perspectives can help students examine problems, understand opportunities as entrepreneurs, create better solutions, and engage in engineering in personally fulfilling ways**. In this book, we advocate for engineering faculty to intentionally incorporate additional perspectives alongside the entrepreneurial mindset into every engineering course, and we provide various examples of doing so. Before we get to our methods, let's first consider what multiple-perspectives-taking offers engineering faculty and students.

L. Bosman et al., *Developing the Entrepreneurial Mindset in Engineering Students through a Perspective-Taking Approach*, Springer Texts in Education,
https://doi.org/10.1007/978-3-032-15123-0_7

Engineering design is a core competency in engineering practice (Murray et al., 2019). To fully understand and solve a problem, engineers must uncover its depth and breadth by considering it from many angles, including those of other stakeholders (e.g., empathetic perspectives) (Surma-aho et al., 2022), and those of other disciplines, such as economic, sustainability, comfort, or safety perspectives (Murray et al., 2019).

Taking up various perspectives supports multiple aspects of engineering design, including problem scoping, design ideation, and design success. Like the Murray et al. (2019) quotation indicates, perspectives-taking or "designers position[ing] themselves" in various stances allows more thorough problem scoping to encourage a wider variety of potential solutions during the design ideation phases. Further, a more solid understanding of the use cases possible for a design can support testing protocols that evaluate the true value or success of a particular design in deployment. If the design requirements and testing results are more closely aligned to the real need, the design is more likely to be successful. Hyysalo (2006) found that incorporating user-identified problems and work-arounds helped designers create better solutions.

Undergraduate engineering students may require more scaffolding of perspectives-taking approaches for several reasons. First, they are novices at this practice. In traditional novice-expert paradigms, **novices tend to see situations as "a compilation of equally relevant parts instead of a complete whole in which only certain parts are relevant"** (Benner, 1984, p. 128). As they learn to evaluate the whole system and understand which parts are relevant, novices increase their discretionary judgment and, little by little, increase their problem-solving expertise.

Second, novice and expert engineers rely on different assumptions and beliefs as they solve problems. **Novice engineers are more often influenced by the emotions accompanying their preexisting beliefs,** and a novice designer's individually held beliefs may be both more rigid and more challenging to overcome (McNeill et al., 2016). Experts also rely on beliefs, but they are more often based on relevant experience instead of feelings. Ahmed et al. (2003) found that expert engineers are more likely to use abstracted intuition from previous engineering situations than novice engineers, who are more likely to call upon one-to-one correlations or single case examples when unpacking problems and ideating solutions. Further, experts are more aware of trade-offs and other limitations of problems and solutions.

Third, novice engineering **students are often presented with limited information** as they approach a design challenge, restricting the field of view that they believe they should be accounting for in their thinking. For example, design briefs may be presented to students, offering them a glimpse of the complicated, ill-defined problems that more expert engineers must tackle (Murray et al., 2019). Design briefs (or similar problem presentations) may be intentionally limited to target specific subject matter. Without deliberate interruption, students are likely to move directly through the few inputs offered in the brief, or connect the problem only to the coursework, thus hampering their development of holistic, transferable problem-scoping and -solving.

7.2 Perspectives-Taking for Student Retention and Belonging

Undergraduate engineering students can become disenchanted with their engineering major if they wait until a capstone course to explore nontechnical perspectives on a problem. For this reason, many engineering programs have incorporated a first-year engineering sequence (e.g., Introduction to Engineering) and design spines in an attempt to keep students engaged. Intentionally incorporating additional perspectives may help keep engineering students in the major by both making them feel more **connected to the humanistic work of engineering** and raising the perceived task value of undergraduate engineering coursework.

Situated expectancy-value theory suggests that **even if students believe they are capable of doing engineering coursework tasks, they may not believe the engineering task itself is valuable.** Perceived task value is defined by four components of value: utility, attainment, intrinsic, and cost. Perceived task value has been shown to align with engagement and emotions during STEM courses, STEM major persistence, and STEM career pursuit (Rosenzweig et al., 2024).

Low perceived task value of engineering coursework may help explain the relatively lower retention and career percentage of women and minoritized races in STEM. There are well-documented gender differences in the value that women place on doing work that contributes to society: Women are more likely than men to prefer work with a clear social purpose (Hill et al., 2010). Taking on user and stakeholder perspectives demonstrates the human-centricity of engineering (Surmaaho et al., 2022), and doing so earlier in the undergraduate education experience may encourage women to persist in engineering, given the focus on human and societal impacts of engineering. Similarly, **Black and Latinx students have been found to value STEM careers that benefit communal interests and are more likely to persist in STEM majors and careers that support them to contribute to communal needs** (Rosenzwig, 2024).

Perspectives-taking can help students who occupy multiple intersecting marginalized identities by validating the complexity of their lived experiences and affirming the importance of viewing engineering coursework and design through multiple lenses. This practice encourages all students, including those in engineering, to explore and value unique viewpoints, aligning with the principles of intersectionality (Cho et al., 2013). Increasing the perceived task value of undergraduate engineering education learning activities and assessments could improve retention of underrepresented minorities in engineering.

For decades, **research has correlated retention in engineering programs with student sense of belonging**, defined as "a student's psychological sense of identification and affiliation with the campus community" (Litzler & Samuelson, 2013, p. 23.674.4). Belonging can be influenced by social and institutional factors such as mentorship, friend circles within and outside of courses of study, course content, racial, gender and socioeconomic minoritization, intersectionality, and self-efficacy (Rainey et al., 2021).

Faculty have the opportunity to influence students' sense of belonging and influence their social and academic integration through perspectives-taking. For example, faculty can support and increase students' sense of belonging by establishing a growth mindset (Hill et al., 2010). Pertinent to this work, Schar et al. (2017) found that curriculum content motivation contributes to academic belonging, a co-component of classroom belonging. **By introducing perspectives that align with students' interests, identities, and motivations, faculty can strengthen connections to content and foster a deeper sense of belonging in the engineering classroom.**

For these reasons and more, **engineering students should be explicitly supported to take on new perspectives and explore their varied influences throughout their undergraduate engineering education, including perspectives that raise the task value of engineering learning activities.** Much undergraduate engineering students' coursework is not directly centered on engineering design challenges; design is often in itself a first-year introductory course or a capstone course during the final year. However, foundational engineering courses can help students understand the value of perspectives-taking long before approaching a full design challenge, naturalizing the process of perspectives-taking for more successful design in the capstone and beyond.

7.3 Taking a New Perspective

Intentionally taking and keeping varied perspectives in mind when solving problems may seem like part of a natural thought process. Some simple, everyday examples are identifying empathetic perspectives, such as "Put yourself in their shoes," and "Do unto others as you'd have them do unto you." Other tools, including pros and cons lists or considering alternatives, can help us all to think through both sides of a debate to justify and rationalize our choices.

There are also several formal methods of perspectives-taking that engage critical thinking. Examples such as systems thinking, cultural-historical activity theory, deBono's six thinking hats, SWOT analysis, theory of constraints, lean thinking, and scenario planning all have various definitions and steps to support thinkers in all kinds of situations. Several of these methods support and connect to our version of perspectives taking, specifically systems thinking, CHAT, and De Bono's six thinking hats. For your reference, the following are brief descriptions of these three methods.

Systems thinking is a set of tools and a framework for analyzing complex "mutually causative" relationships and a language for communicating about "dynamic complexities and interdependencies" (Anderson & Johnson, 1997, p. 17). Anderson and Johnson's "Shape of the Problem" activity (p. 25) centers the problem, then instructs users to make connections to it by asking, "Who else outside your group is directly involved in or affected by this issue?" (p. 26). According to Anderson and Johnson, "When you look at the world systemically, it becomes clear

Table 7.1 The six thinking hats

Hat	Role	Contribution
Blue	Managerial	The big picture
White	Neutral	Facts and information
Red	Intuitive	Feelings and emotions
Black	Pessimist	Negative
Yellow	Optimistic	Positive
Green	Creative	New ideas

that everything is dynamic, complex, and interdependent. Put another way: Things change all the time, life is messy, and everything is connected" (p. 19). Situating the problem within a system of relationships among people and dependencies can help the problem-solver create a more comprehensive solution.

Cultural-historical activity theory (CHAT) is used to describe human mental functions, especially as a contrast between cognition and activity, in order to better understand decisions and causality. CHAT emerged within the field of human psychology from Vygotsky's first-generation activity theory drawing on dialectical materialism and classical German philosophy. Vygotsky's students, Luria and Leont'ev, added "social, cultural, and historical dimensions" to the understanding of human thought. Leont'ev introduced "object-practical activity" (Roth & Lee, 2007, p. 189), meaning that human activities, including learning and problem-solving, can be described and analyzed by considering the dynamics of motivation, societal structures and rules, and the physical means of doing activities.

De Bono's six thinking hats is an application of de Bono's parallel thinking, during which members of a group intentionally focus on separate but related tracks to allow the whole group's processing of a concept or idea to deviate in a controlled way (Aithal & Kumar, 2016). Six thinking hats and parallel thinking remove antagonistic practices in individual, two-party, or group debate. Instead, users of the technique literally or figuratively put on hats associated with six modes of thinking in various specific sequences depending on the challenge at hand (see Table 7.1):

Systems thinking, CHAT, and six thinking hats have the potential to expand a problem space by considering various inputs and impacts to support improved critical thinking skills. In our work, we adapted these approaches within a universal design for learning framework, focusing on creating added value for underrepresented students in engineering. We proceeded with the understanding that strategies designed to support these students would not diminish the experience of white and male students. In fact, research on inclusive teaching practices shows that approaches beneficial to specific groups often prove advantageous for all learners (Florian & Linklater, 2010).

7.4 Our Version of Perspectives-Taking

While there are indeed many well-established frameworks for perspectives-taking, **we've chosen to use a unique subset of perspectives that will capitalize on student interests and values** to provide a contrast to technical norms in core engineering coursework. As previously discussed, in engineering education, the perspectives through which students understand the value of their tasks may be dictated by the course and its content. Cech (2014) identified how such technical priorities create a culture of disengagement in engineering by naming three pervasive pillars of engineering: (1) the depoliticization or framing of social and political issues as irrelevant in engineering, (2) technical-social dualism (i.e., separating social considerations from technical and overvaluing the technical work), and (3) meritocratic idealism, which overlooks system inequalities that contribute to or hinder the primary source of value—individual merit. Together, these pillars form the usual playing field on which engineering students learn what matters in engineering, and they may learn that what they care about does not matter. For instance, a course on programming industrial robots may seem to give primary value to the work of programming, the work that the robots do, and the industry's needs, instead of how robots can be (re) designed to increase accessibility for individuals with physical handicaps, or how autonomous vehicles could improve the safety of workers by reducing lifting injuries.

We want to **help engineering faculty of all courses add value to their engineering coursework tasks** by incorporating perspectives that support the participation and retention of underrepresented students in engineering. Like many of the approaches defined above, our perspectives-taking approach leans on the idea of intentionally shifting from one stance to another, with selective stances to increase student perception of task value.

In our system, the hats or **perspectives called to the fore are bio-inspired design and the incorporation of fine arts into science, technology, engineering, and mathematics (STEAM) to enhance the entrepreneurial mindset.** These perspectives offer direct relevance for students from a wide variety of backgrounds and prior funds of knowledge. We believe that these perspectives offer increased access to engineering coursework, improve perceived task value, and have a high potential impact for equity-based engineering pedagogy.

Our selected perspectives limit the system involved, differentiating it from systems thinking and design thinking, and reducing the cognitive load of defining the whole system. Aligning with CHAT, our perspectives-taking approach recognizes learners' varied social, cultural, and historical experiences and influences, and the learning actions should be constructed and analyzed with these dimensions in mind. We've clarified a few specific hats that we encourage faculty to use as they plan and teach engineering courses. Still, unlike de Bono, we have not defined a sequence for use, thus allowing instructors the flexibility to choose the order in which they introduce perspectives into their course content based on their own teaching expertise and student contexts.

For each perspective, faculty should consider how their curriculum intervention aligns with one aspect of the perspective. Chapter 8 presents the bio-inspired design perspective and the three facets to support faculty in integrating bio-inspired design: emulation, ethical framework, and (re)connection. Chapter 9 unpacks what we mean by STEAM integration and, particularly, faculty options to include artworks, art processes, or art movements in engineering coursework. We encourage faculty to be transparent and open with their students about the perspectives that they are focusing on and why. This will help students to understand what is being asked of them and to internalize the message that there are always multiple perspectives at play in both technical and design-based engineering coursework. The following chapter further explains why we chose to incorporate these unique but related perspectives into engineering learning, as well as the substructures we used to help orient faculty to them.

7.5 Other Perspectives to Consider

While we focused on bio-inspired design and the arts to complement an entrepreneurial mindset, there are clearly many other perspectives and approaches that are worthy of consideration by faculty. We encourage readers to explore other perspective lenses as they develop the skills to integrate more perspectives into engineering courses and to be transparent with their peers and students about why they chose to raise the value of their engineering tasks through taking up specific perspectives. **For your consideration, the following are some additional perspectives that may serve undergraduate engineering well** and increase the perceived task value for underrepresented populations in engineering. The end goal of integrating any of these perspectives is the same as our goal in incorporating the arts and bio-inspired design: a more contextualized and engaging engineering curriculum to help augment students' entrepreneurial mindset development. This list is not exhaustive but intended to support your curiosity or your students'.

Values. Batya Friedman introduced value-sensitive design in 1996 to support technology that "accounts for human values in a principled and systematic manner throughout the design process" (Friedman et al., 2017).

Ethics. Microethics and macroethics in engineering (Herkert, 2001) helps young engineers understand who is ethically responsible in engineered systems: the individual or society.

Generative AI and technology-enhanced learning. Technological mediation theory (Verbeek, 2006) explores how humans and technologies interact from "fusion" interactions, which blur the separation between technology and the body, to immersion interactions, where a technological environment actively interacts with human beings, to augmentation interactions, where technology adds an extra layer of experience (Verbeek, 2015). Verbeek posited that artificial intelligence (AI) is one of many human-technology relations, and technological mediation theory provides us with a language and comparison factors for examining how humans, technologies, and AI interact.

Sustainability. Sustainable design frameworks, such as life cycle assessments, regenerative design, and Engineering for One Planet (Smith et al., 2022) help link engineering with environmental ethics.

Pop culture. The cultural studies approach (Hall, 1980) separates culturalism—the experiences of ordinary people—from structuralism—the imposition of culture and meaning by underlying structures.

Digital civics. Participatory design encourages designers to engage communities in democratic design with digital tools (Bosman et al., 2020; DiSalvo, 2022).

Politics. Science, technology, and society is a vast, multidisciplinary field that affords multiple viewpoints on how science, technology, and society influence one another.

Global perspectives. The global engineering competency framework (Downey et al., 2006) encourages engineers to consider cultural intelligence and collaboration when solving global problems.

Socioeconomic perspectives. Freire's (2020) critical pedagogy remains a powerful tool for examining the intersections among engineering, class, power, and equality. Riley's (2003) work translates Freire to engineering.

Gender equity. Feminist technoscience examines how gender and gender norms influence and are influenced by technology. See Faulkner (2001).

Social justice. Engineering for social justice empowers engineers to meet the justice needs and demands of modern society (Leydens & Lucena, 2017).

Disability studies in engineering and universal design. Disability and accessibility are important considerations for teaching engineering students and for creating engineers with awareness and capacity to support access in their work. Disability studies in engineering (Hamraie, 2016) and universal design for learning (Steinfeld & Maisel, 2012) focus on supporting learners with disabilities, whereas universal design (Burgstahler, 2009) considers the usability of a designed product.

Historical contexts. Engineering historiography explores the interrelations among historical narratives, past decisions, and engineering development. See Winner (2017).

Focusing on human values, ethics, and the morality of stakeholders in engineering design can impact both stakeholder and designer agency, which may increase task value. Connecting engineering learning to real-world issues can provide much-appreciated relevance and grounded applications for helping others.

7.6 You'll Get to the Entrepreneurial Mindset Faster This Way

Integrating perspectives-taking allows instructors the opportunity to more efficiently and effectively develop the entrepreneurial mindset in students. Bio-inspired design and the arts perspectives complement one another and **can help engineering students better understand the holistic and human-centered aspects of engineering and support the development of the entrepreneurial mindset.** Part II of this book explained why and how we should integrate the entrepreneurial mindset into engineering by supporting macro perspectives (discover, evaluate, and exploit*)*, design value (desirability, feasibility, or viability), and micro perspectives (empathize, define, ideate, prototype, and test).

The perspective of bio-inspiration encourages students to discover, evaluate, and exploit naturally successful designs in nature—those that have survived through natural selection and could serve dual purposes in the designed world. Perspectives from the arts, including creating, presenting, responding to, or connecting to artworks, art processes, and art movements, prompts students to understand their value in design and to empathize with stakeholders' needs by learning more about them. Understanding human needs and uncovering opportunities will support students in developing their entrepreneurial mindset.

7.7 Summary

Perspectives-taking supports all individuals, including engineering students, to better understand complex situations by routinizing efforts to consider alternative observations and analyses. In this work, we encourage engineering faculty to incorporate intentional perspectives-taking in their courses to increase the perceived task value of engineering coursework and raise its relevance to students' lives and interests. We recommend that faculty incorporate bio-inspired design and the arts into their engineering coursework and suggest that faculty may find such perspectives support student retention and the development of students' entrepreneurial mindset.

References

Ahmed, S., Wallace, K. M., & Blessing, L. T. (2003). Understanding the differences between how novice and experienced designers approach design tasks. *Research in Engineering Design, 14*, 1–11.

Aithal, P., & Kumar, P. (2016). *Using six thinking hats as a tool for lateral thinking in organizational problem solving*. SSRN.

Anderson, V., & Johnson, L. (1997). *Systems thinking basics*. Pegasus Communications.

Benner, P. (1984). From novice to expert. *Menlo Park, 84*(1480), 10–1097.
Bosman, L., Dautz, E., Van Winkle, R., & Ricco, G. (2020). Integrating digital ciuics into the engineering technology classroom. *Journal of Engineering Technology, 37*(2).
Burgstahler, S. (2009). *Universal design: Process, principles, and applications*. DO-IT.
Cech, E. A. (2014). Culture of disengagement in engineering education? *Science, Technology and Human Values, 39*(1), 42–72.
Cho, S., Crenshaw, K. W., & McCall, L. (2013). Toward a field of intersectionality studies: Theory, applications, and praxis. *Signs: Journal of Women in Culture and Society, 38*(4), 785–810.
DiSalvo, C. (2022). *Design as democratic inquiry: Putting experimental civics into practice*. MIT Press.
Downey, G. L., Lucena, J. C., Moskal, B. M., Parkhurst, R., Bigley, T., Hays, C., Jesiek, B. K., Kelly, L., Miller, J., & Ruff, S. (2006). The globally competent engineer: Working effectively with people who define problems differently. *Journal of Engineering Education, 95*(2), 107–122.
Faulkner, W. (2001). *The technology question in feminism: A view from feminist technology studies*. Women's studies international forum
Florian, L., & Linklater, H. (2010). Preparing teachers for inclusive education: Using inclusive pedagogy to enhance teaching and learning for all. *Cambridge Journal of Education, 40*(4), 369–386.
Freire, P. (2020). Pedagogy of the oppressed. In *Toward a sociology of education* (pp. 374–386). Routledge.
Friedman, B., Hendry, D. G., & Borning, A. (2017). A survey of value sensitive design methods. *Foundations and Trends® in Human–Computer Interaction, 11*(2), 63–125.
Hall, S. (1980). Cultural studies: Two paradigms. *Media, Culture and Society, 2*(1), 57–72.
Hamraie, A. (2016). Universal design and the problem of "post-disability" ideology. *Design and Culture, 8*(3), 285–309.
Herkert, J. R. (2001). Future directions in engineering ethics research: Microethics, macroethics and the role of professional societies. *Science and Engineering Ethics, 7*, 403–414.
Hill, C., Corbett, C., & St Rose, A. (2010). *Why so few? Women in science, technology, engineering, and mathematics*. ERIC.
Hyysalo, S. (2006). The role of learning-by-using in the design of health care technologies: a case study. *The Information Society, 22*(2), 89–99.
Leydens, J. A., & Lucena, J. C. (2017). *Engineering justice: Transforming engineering education and practice*. Wiley.
Litzler, E., & Samuelson, C. (2013). How underrepresented minority engineering students derive a sense of belonging from engineering. *2013 ASEE annual conference & exposition*.
McNeill, N. J., Douglas, E. P., Koro-Ljungberg, M., Therriault, D. J., & Krause, I. (2016). Undergraduate students' beliefs about engineering problem solving. *Journal of Engineering Education, 105*(4), 560–584.
Murray, J. K., Studer, J. A., Daly, S. R., McKilligan, S., & Seifert, C. M. (2019). Design by taking perspectives: How engineers explore problems. *Journal of Engineering Education, 108*(2), 248–275.
Rainey, A. M., Smith, J. M., & Verdin, D. (2021). Classroom practices that support minoritized engineering students' sense of belonging. *2021 ASEE virtual annual conference content access*.
Riley, D. (2003). Employing liberative pedagogies in engineering education. *Journal of Women and Minorities in Science and Engineering, 9*(2).
Rosenzweig, E. Q., Chen, X.-Y., Song, Y., Baldwin, A., Barger, M. M., Cotterell, M. E., Dees, J., Injaian, A. S., Weliweriya, N., & Walker, J. R. (2024). Beyond STEM attrition: Changing career plans within STEM fields in college is associated with lower motivation, certainty, and satisfaction about one's career. International Journal of STEM Education, 11(1), 15.
Roth, W.-M., & Lee, Y.-J. (2007). "Vygotsky's neglected legacy": Cultural-historical activity theory. *Review of Educational Research, 77*(2), 186–232.

Schar, M., Pink, S. L., Powers, K., Piedra, A., Torres, S. A., Chew, K. J., & Sheppard, S. (2017). Classroom belonging and student performance in the introductory engineering classroom. *2017 ASEE annual conference and exposition*.

Smith, V., Shrestha, S., Welker, A., & Sample-Lord, K. (2022). Including principles of sustainability in design by implementing the engineering for one planet framework. *2022 ASEE annual conference and exposition*.

Steinfeld, E., & Maisel, J. (2012). *Universal design: Creating inclusive environments*. Wiley.

Surma-aho, A., Björklund, T., & Hölttä-Otto, K. (2022). User and stakeholder perspective taking in novice design teams. *Design Science, 8*, e24.

Verbeek, P.-P. (2006). Materializing morality: Design ethics and technological mediation. *Science, Technology, and Human Values, 31*(3), 361–380.

Verbeek, P.-P. (2015). COVER STORY beyond interaction: A short introduction to mediation theory. *Interactions, 22*(3), 26–31.

Winner, L. (2017). Do artifacts have politics? In *Computer ethics* (pp. 177–192). Routledge.

Chapter 8
Why Bio-Inspired Design?

Abstract Expanding on the perspectives-taking approach, bio-inspired design is introduced as a powerful framework for fostering creativity, innovation, and systems thinking within engineering education. Bio-inspired design—the process of drawing inspiration from biological systems to develop technologies, products, or services—provides a relevant and engaging way to explore how nature's strategies can support innovation. By using analogy-making, bio-inspired design enables those without a biology background to identify elegant and functional patterns in nature and apply them to human-designed systems. This approach brings a fresh perspective to problem-solving by emphasizing both human needs and ecological insight, inviting transdisciplinary collaboration across science, engineering, and the humanities. This chapter offers an overview of bio-inspired design, highlights its natural connections to engineering disciplines, and provides motivation for its integration into the engineering classroom to encourage broader participation and retention, particularly among students driven by service and societal impact.

Keywords Bio-inspired design · Biomimicry · Transdisciplinary problem-solving

8.1 Learning from Nature

> "What if, every time I started to invent something, I asked 'How would nature solve this?'"
> —*Jane Benyus, cofounder of the Biomimicry Institute*

Bio-inspired design refers to the process of drawing principles, structures, and strategies from biological systems to develop human technologies, products, and services. It is rooted in the idea that natural systems offer tested, efficient, and often elegant solutions to complex challenges (Helms et al., 2009). Unlike conventional design processes that prioritize technical feasibility or market needs in isolation, bio-inspired design invites designers to observe how living organisms and ecosystems adapt, survive, and optimize their functions within their environment.

L. Bosman et al., *Developing the Entrepreneurial Mindset in Engineering Students through a Perspective-Taking Approach*, Springer Texts in Education,
https://doi.org/10.1007/978-3-032-15123-0_8

This approach offers a useful perspective for innovation across a range of fields, including engineering (Yeter et al., 2023). By studying phenomena such as the self-cleaning properties of lotus leaves, the aerodynamic shape of kingfisher beaks, or the temperature regulation of termite mounds, designers can abstract functional principles that inform novel, efficient, and often sustainable human-made systems (Bar-Cohen, 2006).

At its core, **bio-inspired design emphasizes the translation—not replication—of biological strategies into new contexts**. It encourages analogy-making: recognizing patterns or functions in nature and adapting them to solve unrelated human challenges (Nagel et al., 2011). Importantly, bio-inspired design does not require a deep biological background; rather, it is accessible to practitioners across disciplines through structured methods that involve observation, abstraction, and application.

Beyond technological innovation, bio-inspired design also highlights the broader relevance of nature as a guide for addressing human needs (Ilieva et al., 2022). As societies face increasingly complex problems related to sustainability, health, and infrastructure, looking to biological systems offers pathways that integrate functionality, adaptability, and resilience. Thus, bio-inspired design serves as a method for creating novel solutions and rethinking the relationship between human progress and natural systems.

8.2 The Biomimicry Process

Biomimicry is a specific form of bio-inspired design that follows a distinct process for learning from and applying nature's strategies to human challenges (Hayes et al., 2020). **It asks not just what can we learn from nature, but how can we apply these lessons in a way that is sustainable and systems-oriented?**

At the heart of biomimicry is **analogy-making**—translating human problems into functional questions nature has already solved. As Bar-Cohen (2006) notes, biomimetic design begins by identifying biological systems that perform similar functions to the desired outcome, then modeling innovation on those strategies.

Biomimicry reframes innovation not as invention from scratch but as learning from nature's 3.8 billion years of evolution. It encourages solutions that are adaptive, regenerative, and often elegantly efficient. The process unfolds in five key steps, each helping designers approach problems from a nature-informed perspective.

Step 1: Define the challenge. Every design process begins with a problem. But in biomimicry, the problem is framed as a function—*What does the system need to do?* For example: How might we reduce heat in buildings without using air conditioning? How might we transport water with minimal energy?

Step 2: Biologize the question. Next, the design question is translated into nature's language. Instead of asking, "How can we design better insulation?" we

Table 8.1 Example biological strategies and innovations (source: AskNature.org)

Biological Strategies	Innovations
The **forest floor** has a variety of colors and textures, allowing a wide variety of creatures to exist and hide from predators	i2 **carpet tiles** have random patterns and colors, making it difficult to perceive small stains and soil
Evapotranspiration is the natural process by which plants absorb water from the soil and release it into the atmosphere through tiny pores (stomata) in their leaves	changeWATER: Labs has developed a **low-cost, portable toilet** that disposes of human waste by evaporating out the water
As **fish** swim, they shed tiny vortices. When swimming in large schools, individual fish transfer energy to each other with these vortices, lowering the energetic costs of swimming	**Wind turbine** farms from the California Institute of Technology group turbines together to increase energy output
Dogs are expert sniffers, partly because their rapid inhale-exhale cycle pulls distant scents closer to their noses, enhancing detection	The **ventilation system** from the National Institute of Standards and Technology fits onto commercial detectors and both inspires and expires air to improve sampling flow dynamics
Geckos can cling to nearly any surface due to millions of hairlike setae on their toes, which branch into nano-scale spatulae that generate strong Van der Waals forces through their large combined surface area	Magink uses a patented preparation process involving Van der Waals forces to create a **versatile, removable adhesive.**

ask, "How does nature manage temperature?" This shift allows us to look across species and ecosystems for strategies with similar functions.

Step 3: Find nature's best practices. With a biologized question, the next step is to research how organisms and systems achieve this function. Termite mounds regulate temperature through passive airflow. Cacti reduce heat through surface texture. Polar bears trap heat with hollow fur. These examples become models for how we might solve the design problem.

Step 4: Emulate and adapt. Nature's strategies are then translated into design concepts. Engineers and designers study the underlying mechanisms and apply them to new contexts. For instance, the Eastgate Centre in Zimbabwe uses passive cooling inspired by termite mounds. The shape of the kingfisher's beak helped inform the redesign of the front of Japan's bullet train to reduce drag and noise. Additional examples are provided in Table 8.1.

Step 5: Test and refine. Like with any design process, prototyping and feedback are key. Designers test their bio-inspired ideas to evaluate performance, feasibility, and sustainability. The goal is not to copy nature directly but to learn from its principles.

8.3 Connection to Engineering Discipline

Bio-inspired design closely aligns with the core aims of engineering: solving complex problems under real-world constraints. Biological systems that are shaped by evolutionary pressures offer elegant, resource-efficient solutions that mirror the goals of engineering in achieving functionality, adaptability, and sustainability (Yeter et al., 2023). Drawing inspiration from nature encourages engineers to think beyond conventional tools and discover novel approaches grounded in tested ecological principles.

More than a source of design ideas, bio-inspired thinking fosters *systems thinking*, a critical engineering competency (Dugan et al., 2022). Biological systems illustrate how complex interdependencies, feedback loops, and multifunctional designs can create resilient, adaptive outcomes (Bar-Cohen, 2006). Engaging with these systems helps students move beyond linear problem-solving and appreciate the interconnectedness of engineering challenges across contexts.

Incorporating bio-inspired design into engineering also broadens the discipline's relevance. Many nature-based solutions inherently advance sustainability by addressing challenges such as energy efficiency, waste reduction, and environmental resilience (Helms et al., 2009). These connections make engineering feel more purposeful, particularly for students motivated by environmental and societal concerns, and can help attract and retain a more diverse student population.

By encouraging creative exploration, interdisciplinary thinking, and real-world impact, bio-inspired design offers engineering a powerful framework to cultivate both technical excellence and socially conscious innovation (Ilieva et al., 2022).

8.4 Summary

Bio-inspired design offers a compelling perspective for reimagining both technological innovation and engineering education. By studying how nature solves complex challenges, students develop key competencies—such as systems thinking, interdisciplinary collaboration, and purposeful innovation—while also finding deeper relevance in their work. Embedding these principles in engineering education not only equips future engineers to design more resilient and efficient solutions but also fosters inclusion by connecting technical learning to broader social and environmental goals.

References

Bar-Cohen, Y. (2006). Biomimetics—using nature to inspire human innovation. *Bioinspiration and Biomimetics, 1*(1), P1.

Dugan, K. E., Mosyjowski, E. A., Daly, S. R., & Lattuca, L. R. (2022). Systems thinking assessments in engineering: A systematic literature review. *Systems Research and Behavioral Science, 39*(4), 840–866.

Hayes, S., Desha, C., & Baumeister, D. (2020). Learning from nature–Biomimicry innovation to support infrastructure sustainability and resilience. *Technological Forecasting and Social Change, 161*, 120287.

Helms, M., Vattam, S. S., & Goel, A. K. (2009). Biologically inspired design: Process and products. *Design Studies, 30*(5), 606–622.

Ilieva, L., Ursano, I., Traista, L., Hoffmann, B., & Dahy, H. (2022). Biomimicry as a sustainable design methodology—Introducing the 'Biomimicry for Sustainability'framework. *Biomimetics, 7*(2), 37.

Nagel, J. K., Nagel, R. L., & Stone, R. B. (2011). Abstracting biology for engineering design. *International Journal of Design Engineering, 4*(1), 23–40.

Yeter, I. H., Tan, V. S. Q., & Le Ferrand, H. (2023). Conceptualization of biomimicry in engineering context among undergraduate and high school students: An international interdisciplinary exploration. *Biomimetics, 8*(1), 125.

Chapter 9
Why STEAM (in Particular, the Arts)?

Abstract Continuing this perspectives-taking exploration, incorporating the arts into engineering education is posited to nurture empathy, creativity, and the holistic problem-solving essential for the engineers of the future. Arts integration in STEM, or STEAM, refers to the inclusion of fine arts, specifically visual, theatrical, dance, and music arts, in conjunction with other core STEM content learning. Arts integration can support STEM discipline learning by fostering relevance, capturing student interest, activating and reifying neural networks to call in multiple intelligences, and constructing a deeper understanding of both the arts and core content. The National Core Arts Standards provide guidance for art instruction, which can be included as learning objectives within other courses to create an integrated lesson, assignment, or assessment. This chapter provides an overview of STEAM, highlighting its natural connections to various engineering disciplines, and advocates for its use in the engineering classroom to drive innovation and problem-solving. Drawing from diverse sources of inspiration, including art, literature, music, and nature, engineers can generate innovative ideas and approaches that transcend traditional boundaries.

Keywords STEAM · Retention · Arts standards · Arts integration

9.1 What's Art Got to Do with It?

> "Students who have arts-rich experiences at school do better across the board academically, and they also become more active and engaged citizens."
> —*Catterall et al.* (2012, p. 3)

Once upon a time, the arts were integrated naturally and without fuss into our learning lives. As children, songs helped us learn and remember all kinds of new information, such as routines ("Clean Up Song"), history ("Ring Around the Rosie"), or civics ("I'm Just a Bill"). Illustrations in stories inspired our mind's eye, and drawing from imagination provided expression with the simplest instruments: crayons and paper. Multiple representations followed us to higher math, where

L. Bosman et al., *Developing the Entrepreneurial Mindset in Engineering Students through a Perspective-Taking Approach*, Springer Texts in Education,
https://doi.org/10.1007/978-3-032-15123-0_9

equations are equally valid as text and as graphical representations, but the art was removed. And yet, when mathematicians, scientists, or engineers preserve art in their practice, be it in song, creative storytelling, or visual expressions of ideas or inventions, they are venerated. For example, the works of Da Vinci and Audubon clearly demonstrate visual art overlapping with engineering and science. Feynman's lectures, like a precursor to TED Talks, provide rich storytelling experiences to explain fundamental physics and its paradoxes. Moreover, some artists were engineers first, including Calder and Kandinsky.

Research on arts integration in education settings shows that curricular and extracurricular arts education correlates to higher academic achievement. Analysis of four longitudinal data sets collected by the US Department of Education (NELS:88, ECLS-K, and ELS:2002) and the US Department of Labor (NLSY97) found that for students in the lowest quartile of socioeconomic status (SES), those with high arts-rich experiences were 23% more likely to attend college after high school, and more than twice as likely to attend a 4-year college after high school. If these students attended college, those in the lowest quartile of SES with **high arts-rich experiences were three times as likely to graduate with a bachelor's degree** (Catterall, 2012). Students who took four or more years of arts and music classes averaged almost 100 points higher on their SAT scores versus students who took a half year or less (Engebretsen et al., 2013). Additional findings revealed more statistically significant correlations for this population; high arts-rich experiences correlated to:

- Higher eighth-grade science test scores, and higher eighth-grade writing test scores;
- A higher rate of tenth-graders who went on to complete a high school calculus course;
- Higher high school mean GPA (the same trend was found among high SES students), and higher high school mean math GPA;
- Five times higher high school graduation rate;
- Higher intention to earn a bachelor's degree among eighth- and twelfth-graders; and
- **Higher intention to choose professional majors and careers** (three times more likely than students with fewer arts-rich experiences to choose nursing/allied health or biology majors, and three times more likely to choose nursing or other medical professional careers; data aggregated for engineering was not made available).

Arts instruction, both curricular and extracurricular, is good! In addition to the academic and retention correlations listed above, **art-making can support cognitive, attitudinal, behavioral, health and wellness, and social and economic improvements** (Weiss, 2023). But, sadly, since the 1980s, the National Endowment for the Arts (NEA), which, along with the National Assembly of State Arts Agencies, funds and promotes arts opportunities throughout the country, has been chronically underfunded. According to the non-profit arts advocacy group Americans for the Arts, if government arts funding had maintained its 1984 percentage of nondefense discretionary spending—0.107%—then the 2018 NEA budget would have been $771.5 million instead of only $152.8 million (Cohen, 2019). In 2022, total federal, state, and local funding for the arts was $1.85 billion, or $5.49 per American (Stubbs & Mullaney-Loss, 2020).

Even with reduced overall funding, arts instruction is still present throughout the United States: **all 50 states, the District of Columbia, and the Department of Defense Education Activity have arts education standards at the early childhood, elementary, and secondary level** (Arts Education Partnership, n.d.). Specific requirements for number of classroom hours per week and number of courses for graduation vary by state. The critical message for engineering faculty is that **all your students, even the most technically and mathematically minded ones, have had years of arts education, ranging from classes in the visual arts, music, dramatic theater, and/or dance.** The 2019 NAEP High School Transcript Survey (National Assessment of Educational Progress (NAEP), 2020) revealed that on average, high school graduates took 2.3 Carnegie Unit equivalents in visual and performing arts. Each Carnegie Unit (or credit) is equivalent to a year-long course, or 120 hours. For comparison, the same group took an average of 2.2 Carnegie Units of foreign language and 4.0 Carnegie Units of math. Students entering college, including engineering programs, have a wealth of arts experience that engineering faculty can draw upon using an asset-based instructional approach. Why ignore it?

9.2 Why Has STEAM Become a Buzzword, and What Evidence Is There for STEAM Success?

The acronym STEM became popularized as a curricular and programmatic emphasis and opportunity to increase the United States' technical ferocity in the global marketplace by improving STEM education after the National Academies (2007) published *Rising Above the Gathering Storm* identified a gap in American preparedness in science and mathematics fields. Elementary, secondary, and tertiary STEM programs and schools, including charter schools, have been growing for decades. From 2005 to 2015, the number of STEM schools exploded (Lynch et al., 2013). Bipartisan political support and funding for STEM education was supported by reports on how STEM will enhance national interests and global competitiveness. The President's Council of Advisors on Science and Technology, 2010 report noted that "the United States cannot remain at the forefront of science and technology if the majority of its students—in particular women and minorities underrepresented in STEM fields—view science and technology as uninteresting, too difficult, or closed off to them" (President's Council of Advisors on Science and Technology, 2010). Interestingly, although STEM adoption has grown in education, some studies show that around 50% of US adults outside of education fields could not name the four component parts of the acronym accurately (Pew Research Center, 2018; Rotermund & Burke, 2021). This indicates that, without direct STEM education expertise and experience, most Americans are underinformed about STEM.

The use of the arts in STEM is a comparatively newer endeavor. STEAM— science, technology, engineering, arts, and mathematics—became a popular sidecar to the STEM conversation around 2010. The first grant for a childhood STEM Learning Through the Arts initiative was awarded in 2010 to the Wolf Trap Foundation for the performing arts. In the last 20 years, STEM has become nearly ubiquitously understood, and yet STEAM remains a bit more mysterious. According to Aguilera and Ortiz-Revilla (2021), the *A* was introduced to emphasize and encourage creativity in STEM. Their literature review indicates, however, that there are many and contradictory forms of STEM and STEAM, both in theory and in practice. They also found that regarding creativity, both STEM and STEAM positively impact student creativity.

The Institute for Arts Integration and STEAM emphasizes that STEAM includes the natural inquiry and problem-solving found in science and engineering, respectively. According to the Institute, "STEAM Education is an approach to learning that uses Science, Technology, Engineering, the Arts and Mathematics as access points for guiding student inquiry, dialogue, and critical thinking" (Riley, 2021). In this book, we refer to STEAM as "STEM with an emphasis on the arts" and expect that our audience of educators will have a preexisting understanding of STEM education.

Land (2013) contends that STEAM learning can improve innovation in STEM, a welcome result if the goal is global competitiveness. And yet, STEAM is useful for more than just beating other countries to the market; by emphasizing the arts, STEAM can engage engineering students in ways that traditional STEM coursework may not, support their core content achievement (Graham & Brouillette, 2016; Inoa et al., 2014), foster critical and creative thinking (Harris & De Bruin, 2018), and encourage personal reflection and metacognition (Cunningham et al., 2015).

9.3 When Is STEAM Truly Integrated?

During the past 20 years, copious research on STEM education has enumerated various methods of STEM integration from full integration—i.e., interdependence between subject content, or the need for each subject to be used to teach other subjects—to a siloed approach, in which each subject in STEM or STEAM is taught independently but is indeed taught (Bybee, 2015). The following spectrum of integration (Fig. 9.1) describes the various ways to connect multiple subjects in interdisciplinary education. This figure was adapted by Shirey based on Riley (2018) and informed by co-teaching and co-planning experience. Shirey has found that to effectively align varying disciplinary content, it is often helpful to consider multiple potential ways to work together in planning, construction of lessons, activities, and assessments, and to provide options for teaching the integrated content based on student and educator availability, time, and level of support needed. The idea of integration is rooted in building curricular coherence to support building

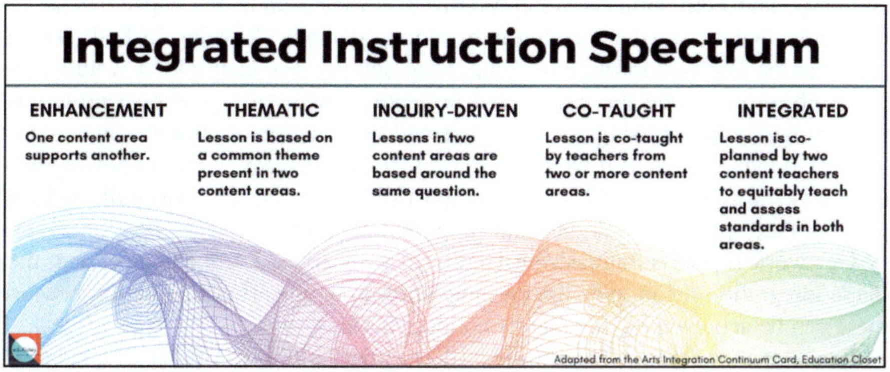

Fig. 9.1 The integrated instruction spectrum (source: Shirey, 2025)

knowledge, resting on a history of research in cognitive science and student learning outcomes (Beane, 1997; Fogarty & Stoehr, 2008; Tomlinson & McTighe, 2006).

Enhancement allows educators to explore the early stages of highlighting cross-curricular or interdisciplinary concepts with little to no need to co-plan with another educator. In engineering, this could be achieved by mentioning the mathematical content from a mathematics class that can support students in a statics class, for example.

Thematic instruction allows two collaborating educators to share disparate disciplinary perspectives on the same theme with the same group of students, drawing out how the disciplines are unique and how they relate. For example, educators in the mathematics department and engineering department who share students in the same term could both center some of their instructional lectures, applications, examples, and assessments on the same theme, such as "sewage treatment in mid-sized U.S. cities."

Inquiry-driven integration moves beyond thematic awareness to the higher orders of thinking in coordinated courses' instruction based around one question, turning our prior theme into, "How efficient is sewage treatment in the largest neighboring city and why? What inputs and outputs contribute to our understanding of efficiency? How has that efficiency changed over time, and what are the projections for the future?" Separately, students could learn from faculty from different departments how to address the questions from their own disciplines, while students note that information in one discipline relates to that in another.

In a co-taught lesson, teachers work together with the students at the same time. They may choose how to sequence their instruction so that both perspectives are explored and given adequate time and emphasis in the classroom. A benefit to co-taught instruction is that when a discipline-specific question arises, both educators are available to share their insights and expertise on the spot.

Finally, integrated instruction refers to a co-planned instruction that straddles two or more disciplinary areas, and in which the disciplines are both (or all) necessary to complete the instructional activity. Therefore, the material is treated equitably, and both subjects are emphasized in their standalone capacities and in their contributions to the learning experience. Students' learning in both subject areas can be assessed through one integrated activity. For example, it would be as impossible to evaluate the causes and projections of a city's compost programs' effectiveness without being able to understand the function of microbes in compost to break down detritus as it would be to evaluate the efficacy without knowing how energy is transferred outside of the compost pile, how the data was collected and whether there were technologies that supported the effort.

Our work supports engineering faculty to work toward arts enhancement, thematic integration, or inquiry-driven integration in engineering instruction because these can be achieved by one faculty member in one course.

Lynch et al. (2013, p. 5) identified 10 essential hallmarks of STEM education, which have been studied extensively since 2005. Our work in this book focuses on eight hallmarks:

- STEM-focused curriculum;
- Reformed instructional strategies and project-based learning for active teaching and learning, including assessment;
- Integrated and innovative technology use;
- Real-world STEM partnerships;
- Early college-level coursework;
- Well-prepared STEM teaching staff;
- Inclusive STEM mission; and
- Support for underrepresented students.

9.4 How Could You Integrate the Arts into a Core Subject?

A fantastic way to integrate the arts into other content areas is to lean on existing arts learning standards. The National Coalition for Arts Standards (2014) provides guidance to teachers through learning standards on what students should learn in arts education from kindergarten through grade 12.[1] Although the National Core Arts Standards (NCAS) are intended for K-12 audiences, their structure and specific learning tasks can be instructive and inspiring for all educators. First, the NCAS can help engineering faculty define what counts as art for arts integration and STEAM simply by what is included in the standards: learning targets for students in visual arts, media arts, instrumental music, music composition, music technology, theater, and dance.

Second, the NCAS separate out four aspects of arts learning for each genre: creating, performing and presenting, responding, and connecting. With these

[1] Since the 1990s, use of "learning standards" has been a common method to instruct teachers on what they should teach their students Reigeluth, C. M. (1997). Educational standards. *Phi delta kappan, 79*(3), 202-206.

distinctions, the NCAS make it clear that learning art does not only mean making something tangible or performing it. Learning art can also mean connecting art to one's own lived experiences, or responding to artworks through selection, analysis, interpretation, or evaluation. For example, as noted in NCAS standard MU:Re9.1.T.IIa, students who choose lead-in music for a podcast can "develop and justify the evaluation of a variety of music based on established and personally-developed criteria, digital, electronic and analog features, and understanding of purpose and context" (Music Tech, National Coalition for Core Arts Standards [2014] National Core Arts Standards. Rights Administered by the State Education Agency Directors of Arts Education. Dover, DE, p. 3).

Faculty have choices when it comes to integrating the arts into their coursework: Which medium or genre will I focus on (visual arts, theater, and so on), and what aspect of arts interaction will I focus on (creating, performing and presenting, responding, or connecting)? From working with educators who are not trained artists, we found that it can be useful to specifically focus on one artwork, one art-making process, or one art movement to help narrow the breadth of options and to clarify what content to present to students.

An example of an integrated task for engineering students is provided in Table 9.1 for each intersection. Please note that these could be used with any second perspective; however, Table 9.1 emphasizes intersections specifically with a bio-inspired design perspective.

9.5 Summary

Arts integration in STEM—or STEAM—can help engineering faculty bring student interests, creativity, and productivity into the engineering classroom, disrupting the idea that engineering is purely pragmatic and objective. Although STEM is the more prominent acronym, both STEM and STEAM offer positive impacts on student learning. Arts integration can take various forms. Leaning on national arts learning standards can support engineering faculty in integrating artworks, art processes, and art movements into their courses.

Table 9.1 Examples of arts-integrated activities (National Coalition for Arts Standards, 2014) aligned to artworks, art processes, and art movements

Student Activity	Artworks	Art Processes	Art Movements
Creating	Compose and improvise ideas for arrangements, sections, and short compositions that reflect characteristic(s) of music referencing nature or animals from a variety of cultures studied in rehearsal (MU:Cr1.1.E.IIa)	Choose from a range of natural materials and methods of traditional and contemporary artistic practices to plan student designs. (VA:Cr1.2.IIa)	Investigate historical and cultural conventions related to the depiction of specific natural phenomenon (i.e., floods, earthquakes, dust storms, tornados) and their impact on the visual composition of a drama/ theatre work. (TH:Cr1.1.II.a)
Performing and presenting	When analyzing selected music inspired by natural systems, read and perform the music using iconic and/or standard notation. (MU:Pr4.2.4)	Initiate, plan, and direct rehearsals with attention to technical details and fulfilling artistic expression. Use a range of rehearsal strategies to achieve performance excellence. (DA:Pr5.1.III)	Perform with expression and technical accuracy, in individual and small group performances, a varied repertoire of music that includes melodies, repertoire pieces, improvisations, and chordal accompaniments in a variety of styles, demonstrating sensitivity to the audience and an understanding of the context (natural, social, cultural, and historical). (MU:Pr6.1.H.IIa)
Responding	Identify and interpret works of art or design that reveal how people live around the world and what they value, specifically as related to representations of bio-inspired design. (VA:Re.7.1.6a)	Use bio-inspired thinking to consider and justify the aesthetics of the production elements in a drama/theatre work. (TH:Re9.1.I)	Develop, justify and defend interpretations of varied works, demonstrating an understanding of the composers' intent by citing the use of elements of music (including form), compositional techniques, and the style/genre and context of each work, including regional or global natural inspirations. (MU:Re8.1.C.IIIa)

(continued)

Table 9.1 (continued)

Student Activity	Artworks	Art Processes	Art Movements
Connecting	Compare two dances with contrasting themes. Discuss feelings and ideas evoked by each. Describe how the themes and movements relate to points of view and experiences, consider connections to natural forms and movements. (DA:Cn10.1.5)	Analyze and responsibly interact with media arts tools and environments, considering copyright, ethics, sustainability, media literacy, and social media. Consider bio-inspiration to mitigate concerns (MA:Cn11.1.7b)	Describe how knowledge of culture, traditions, and history, especially as related to biological influence, science, and or nature may influence personal responses to art. (VA:Cn11.1.Ia)

References

Aguilera, D., & Ortiz-Revilla, J. (2021). STEM vs. STEAM education and student creativity: A systematic literature review. *Education Sciences, 11*(7), 331.

Arts Education Partnership. (n.d.) *ArtScan At-a-Glance*. Retrieved 9 Apr. 2025, www.aep-arts.org/artscan/artscan-at-a-glance/

Beane, J. A. (1997). *Curriculum integration: Designing the core of democratic education*. Teachers College Press.

Bybee, R. W. (2015). *The case for STEM education: Challenges and opportunities. In The case for STEM education: Challenges and opportunities*. National Science Teachers Association. doi:https://doi.org/10.2505/9781936959259.

Catterall, J. S. (2012). *The arts and achievement in at-risk youth: Findings from four longitudinal studies*. Research Report# 55. National Endowment for the Arts.

Cohen, R. (2019). *NEA budget losses*. Retrieved 5/4/25 from https://www.americansforthearts.org/by-program/reports-and-data/legislation-policy/naappd/nea-budget-losses

Cunningham, P., Matusovich, H. M., Hunter, D. A. N., McCord, R. E., & Ieee. (2015). Teaching metacognition: Helping engineering students take ownership of their own learning. In *Frontiers in education conference* (pp. 367–371). IEEE. <Go to ISI>://WOS:000371705200065.

Engebretsen, K., Van Fleet, E., & Mazzocco, A. (2013). Arts education navigator: Facts and figures. *Americans for the Arts, 1*.

Fogarty, R. J., & Stoehr, J. (2008). *Integrating curricula with multiple intelligences: Teams, themes, and threads*. Corwin Press.

Graham, N. J., & Brouillette, L. (2016). Using arts integration to make science learning memorable in the upper elementary grades: A quasi-experimental study. *Journal for Learning through the Arts, 12*(1).

Harris, A., & De Bruin, L. R. (2018). Secondary school creativity, teacher practice and STEAM education: An international study. *Journal of Educational Change, 19*, 153–179.

Inoa, R., Weltsek, G., & Tabone, C. (2014). A study on the relationsh ip between theater arts and student literacy and mathematics achievement. *Journal for Learning Through the Arts, 10*(1).

Land, M. H. (2013). Full STEAM ahead: The benefits of integrating the arts into STEM. *Procedia Computer Science, 20*, 547–552.

Lynch, S. J., Behrend, T., Burton, E. P., & Means, B. (2013). Inclusive STEM-focused high schools: STEM education policy and opportunity structures. In *Annual conference of national association for research in science teaching (NARST)*. Rio Grande.

National Assessment of Educational Progress (NAEP). (2020). *The Nation's report card: Results from the 2019 NAEP High School Transcript Study*. Retrieved 5/4/2025 from www.nationsreportcard.gov/hstsreport/#coursetaking

National Coalition for Core Arts Standards. (2014). *National core art standards*.

Pew Research Center. (2018). *What Americans know about STEM: Science, technology, engineering and mathematics*. https://www.pewresearch.org/science/2018/01/09/the-public-views-on-stem-education/

President's Council of Advisors on Science and Technology. (2010). *Prepare and inspire: K-12 education in science, technology, engineering, and math (STEM) for America's future*. Doi:https://doi.org/10.1007/s00115-009-2682-8.

Reigeluth, C. M. (1997). Educational standards. *Phi delta kappan, 79*(3), 202–206.

Riley, S. (2018). *5 Shades of integration*. The Institute for Arts Integration and STEAM. https://artsintegration.com/2013/01/25/5-shades-of-integration/

Riley, S. (2021). *What is STEAM education*. The Institute for Arts Integration and STEAM. https://artsintegration.com/what-is-steam-education-in-k-12-schools

Rotermund, S., & Burke, A. (2021). *Elementary and secondary STEM education. Science & engineering indicators 2022. NSB-2021-1*. National Science Foundation.

Shirey, K. (2025). *What does "Integrated Instruction" mean to you?* www.edukatey.com, Retrieved 05/24/2025 from https://www.edukatey.com/videos-and-blogs/#Spectrum

Stubbs, R., & Mullaney-Loss, P. (2020). Public funding for the arts, 2020. *Public Funding for Arts and Culture, 31*(3), 9–12.

Tomlinson, C. A., & McTighe, J. (2006). *Integrating differentiated instruction & understanding by design: Connecting content and kids*. Ascd.

Weiss, E. D. (2023). *Funding for arts in education*.

Part IV
Practice: A Guide for Engineering Educators

Chapter 10
From Theory to Practice: Applying the Perspectives-Taking Approach

Abstract Building on the perspectives-taking approach presented in Part III, this chapter shifts from *understanding* the approach to actively *applying* it in practice through the Scholarship of Teaching and Learning (SoTL). SoTL is a is a field of academic inquiry and practice focused on researching, understanding, and improving teaching and learning in higher education. SoTL encompasses a wide range of activities, including conducting research on teaching methods, assessing student learning outcomes, exploring the effectiveness of educational technologies, and investigating the impact of various instructional strategies on student engagement and success. Engaging in SoTL involves rigorous inquiry grounded in disciplinary knowledge and pedagogical theory. Faculty members, instructional designers, educational researchers, and other professionals in higher education engage in SoTL to enhance their teaching practices, contribute to the scholarship of their disciplines, and promote evidence-based approaches to teaching and learning. This chapter provides an overview of the SoTL Accelerator, a free online resource available at https://www.sotlaccelerator.com/, which guides engineering educators through the SoTL process of curriculum development, assessment, and dissemination.

Keywords Scholarship of teaching and learning (SoTL) · Assessment · Dissemination · IRB · Backward course design · Evidence-based practices

10.1 Scholarship of Teaching and Learning (SoTL)

> "Teaching is not just a transfer of knowledge, but a collaborative process of inquiry and discovery."
>
> —*Pat Hutchings, Scholar and Senior Fellow of the Australian Museum*

SoTL refers to the rigorous and systematic study of teaching practices and student learning with the goal of improving educational outcomes (Gansemer-Topf et al., 2024). **Rooted in reflective practice, SoTL encourages educators to treat their classrooms as sites for scholarly inquiry**, using evidence-based methods to

L. Bosman et al., *Developing the Entrepreneurial Mindset in Engineering Students through a Perspective-Taking Approach*, Springer Texts in Education,
https://doi.org/10.1007/978-3-032-15123-0_10

investigate questions about learning and instruction (Kern et al., 2015). Faculty engaged in SoTL typically collect and analyze data on student performance, attitudes, or engagement, and share their findings through presentations, publications, or campus initiatives.

The **benefits of SoTL** are significant: It fosters continuous improvement in teaching, promotes deeper student learning, strengthens interdisciplinary collaboration, and contributes to a culture of evidence-based education (Gilpin & Liston, 2009). Additionally, it often provides a pathway for faculty to integrate their teaching and research responsibilities more meaningfully.

However, there are **notable challenges**. Engaging in SoTL requires time, methodological expertise, and, often, institutional support, which can be difficult to secure in environments where traditional disciplinary research is more highly valued (Boshier, 2009). Furthermore, translating classroom-based findings into broadly applicable insights can be complex, particularly when dealing with diverse student populations and instructional contexts (Fukuzawa et al., 2020). Despite these hurdles, SoTL has become an essential component of engineering academic life (Rios et al., 2025; Rüütmann & Saar, 2017), encouraging educators to advance not only their own teaching but also the collective knowledge of how students learn best.

SoTL is especially important for the professional development of engineering educators because it bridges the gap between technical expertise and effective teaching. In engineering, where complex problem-solving, critical thinking, and hands-on skills are essential, traditional lecture-based methods may not always meet the needs of diverse learners.

Yet **most new PhD graduates are often not fully prepared to teach** in academia because their doctoral training overwhelmingly focuses on developing research expertise rather than teaching skills (Buswell, 2021). In most PhD programs, the primary goal is to produce original research and contribute new knowledge to a field, and success is usually measured by publications, grants, and technical contributions. As a result**, formal training in pedagogy, curriculum design, assessment methods, and classroom management is often minimal or completely absent.** Even when teaching assistantships are part of a PhD program, they typically involve grading, tutoring, or assisting with labs rather than providing comprehensive instruction or mentorship in how to design and deliver effective courses (Garcia & Kittur, 2024).

In summary, conducting SoTL research can be particularly challenging for engineering educators for several reasons. **First**, many engineering faculty are primarily trained in technical research methods, not educational research, which often involves different methodologies such as qualitative analysis, mixed methods, or classroom-based experimentation. Adapting to these approaches requires additional learning and skill development. **Second**, engineering educators may encounter challenges in generalizing their findings, as classroom innovations that work in one technical course or institution may not easily transfer to others with different student populations or curricular structures. Despite these challenges, **many engineering educators find SoTL valuable for improving student outcomes** and enhancing their own teaching practices.

Moreover, a solution exists! Keep reading.

10.2 Applying Perspectives-Taking with the SoTL Accelerator

The previous three chapters (7–9) presented the perspectives-taking approach, highlighting how integrating the entrepreneurial mindset alongside diverse viewpoints, such as bio-inspired design and the arts, can enhance engineering education and better prepare students for professional practice. Building on this conceptual foundation, the current chapter focuses on translating perspectives-taking into actionable teaching strategies through the Scholarship of Teaching and Learning (SoTL).

The SoTL Accelerator, established in 2021 with funding from the Kern Family Foundation, is a free online resource available at https://www.sotlaccelerator.com/ that guides engineering educators through the SoTL process of curriculum development, assessment, and dissemination. The SoTL Accelerator can be summarized with three key phases associated with the new curriculum development process:

- Phase 1: Develop and Implement
- Phase 2: Assess and Reflect
- Phase 3: Disseminate Findings

10.3 SoTL Accelerator - Phase 1: Develop and Implement

Backward course design is a planning approach for developing and implementing educational experiences that starts by identifying the desired learning outcomes before designing assessments and instructional activities (Wiggins & McTighe, 2005). Instead of beginning with lectures or content coverage, instructors first ask: *What should students know, be able to do, or value by the end of this course?* Once the learning goals are clearly defined, the next step is to design assessments (such as exams, projects, presentations, or other tasks) that provide evidence that students have achieved those goals. Only after setting outcomes and assessments do instructors plan the specific lessons, readings, and activities that will best prepare students for success. This method ensures that every part of the course is aligned and purposeful, leading to more coherent and effective instruction. Backward course design is especially helpful in engineering and other technical fields because it encourages instructors to focus on the development of key competencies, critical thinking, and application of knowledge, rather than just covering large amounts of material.

A key resource in the assess and reflect process is the **SoTL Accelerator Tool #1: Teaching Blueprint** (https://www.sotlaccelerator.com/our-approach), shown in Fig. 10.1, which provides a visual framework to guide backward course design by aligning learning goals, objectives, activities, and assessments. It emphasizes key elements essential to effective curriculum development: (1) the primary teaching strategy, such as integrating entrepreneurial thinking, STEAM concepts, and bio-inspired design; (2) the advancement of professional skills, particularly in communication and teamwork; (3) the encouragement of mindset growth through repeated engagement and feedback to reinforce habits; and (4) the application of best practices in teaching and learning.

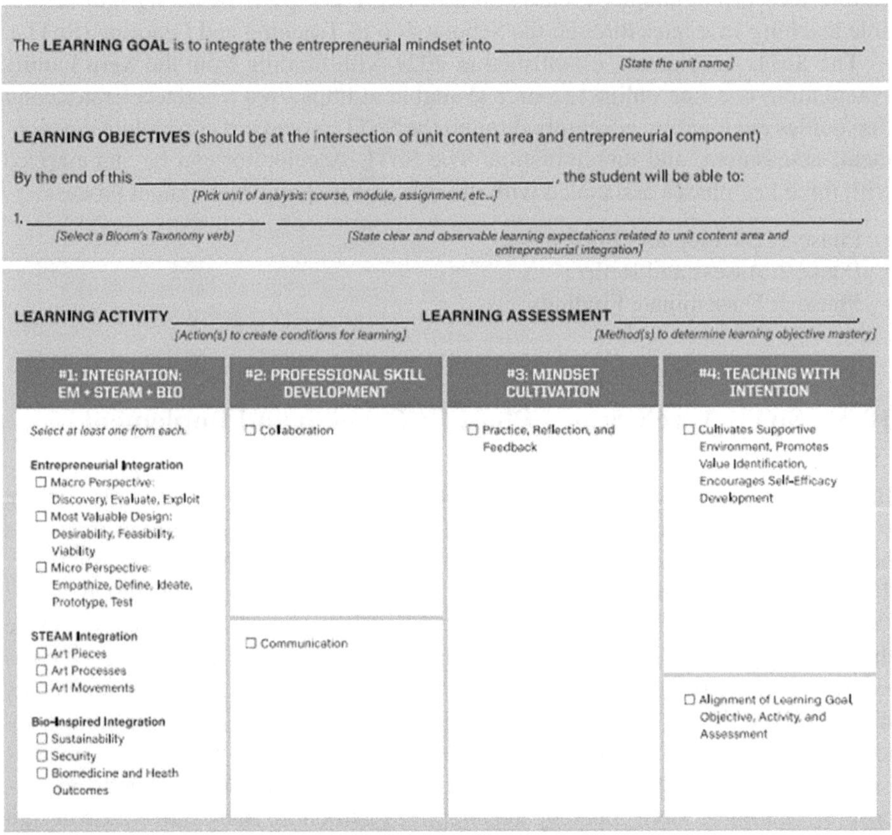

Fig. 10.1 Teaching Blueprint [Modified from Bosman and Fernhaber (2018)]

10.4 SoTL Accelerator - Phase 2: Assess and Reflect

Assessing a SoTL intervention presents several challenges, particularly when trying to measure the effectiveness of teaching strategies in enhancing student learning (Dickson & Treml, 2013; Kennedy et al., 2020). One major challenge is determining appropriate assessment methods that align with the specific learning outcomes of the intervention. In many cases, these outcomes may go beyond traditional knowledge acquisition to include skills such as critical thinking, collaboration, or creativity, which are harder to quantify. Additionally, the diversity of student backgrounds, learning styles, and prior knowledge complicates the task of creating assessments that are both fair and meaningful for all participants. The influence of external factors such as class size, course format, or institutional context can also affect the intervention's success, making it difficult to attribute changes in student performance directly to the intervention itself. Moreover, the iterative nature of SoTL means that assessments need to be continuously refined and adjusted based on feedback, adding further complexity to the evaluation process. Finally, gathering valid and reliable data (whether through surveys, student work, or observations) often requires careful planning and additional time, making it challenging for faculty to balance these tasks with their other teaching and research responsibilities.

Qualitative assessments One approach for assessment that typically works well with smaller class sizes and aims to explore the nature of an inquiry is the use of qualitative open-ended reflection prompts (Rahimi & Maloku-Morina, 2023; Wiliam, 2017). Qualitative reflections offer significant benefits for assessing a SoTL intervention, particularly because they provide rich, detailed insights into the student learning experience that quantitative data alone may miss. Through reflective writing, interviews, or open-ended surveys, students can articulate how they engaged with course materials, how their understanding evolved, and how they perceived the effectiveness of specific teaching strategies. These reflections often reveal nuances in student thinking, motivation, and challenges that are not easily captured through traditional exams or numerical surveys. By analyzing patterns in student reflections, educators can better understand the strengths and limitations of their intervention, identify unexpected outcomes, and adjust their teaching approaches more effectively. Additionally, qualitative reflections empower students to take an active role in the learning process by encouraging metacognition, thinking about how they learn, which can enhance their educational experience. Overall, integrating qualitative reflections into SoTL assessment enriches the evaluation process, leading to more holistic, meaningful, and actionable insights into teaching and learning.

Quantitative assessments Another approach for assessment that typically works well with larger class sizes and aims to explain the nature of an inquiry is the use of quantitative retrospective post-then-pre surveys (Bhanji et al., 2012; Coulter, 2012). A quantitative retrospective post-then-pre survey offers several important benefits for assessing a SoTL intervention, especially when trying to capture shifts in students' knowledge, skills, or attitudes over time. Through this approach, students complete a single survey at the end of the intervention, rating their current abilities and retrospectively assessing their prior abilities before the learning experience. One key advantage is that it minimizes response-shift bias, whereby students often do not fully understand what they don't know at the beginning of a course, so their initial self-assessments can be inaccurate. By reflecting after the experience, students can more accurately judge their growth and learning. This method also streamlines data collection, reducing survey fatigue since it only requires one administration instead of two. Furthermore, the quantitative nature of the survey produces clear, analyzable data that can be used to demonstrate statistically significant learning gains, support program evaluation, and justify instructional improvements. Overall, retrospective post-then-pre surveys provide a practical, efficient, and insightful way to measure the true impact of teaching interventions on student development.

IRB and human subjects research The Institutional Review Board (IRB) is a committee established at research institutions to review and oversee research involving human subjects, ensuring that studies are conducted ethically and that participants' rights, welfare, and privacy are protected (Linder et al., 2014). New engineering PhD graduates might not be familiar with the IRB process and human subjects research because their doctoral training typically focuses on technical or laboratory-based research, rather than studies involving people. In most engineering fields, research projects often involve materials, machines, simulations, or purely quantitative data, where ethical oversight for human subjects is not usually required. As a result, engineering PhD programs often do not include formal instruction on research ethics related to human participants, informed consent, confidentiality, or IRB procedures.

Furthermore, if graduate students do encounter human subjects in their work, such as through user testing or surveys, they may do so under the supervision of a principal investigator who handles the IRB process, leaving the students with little direct exposure to it. This gap becomes especially noticeable when new engineering faculty engage in SoTL or educational research, where understanding and complying with human subjects protections becomes critical. Without prior training, navigating the IRB process can seem confusing or burdensome for those transitioning into roles that combine technical and educational research responsibilities.

In the context of SoTL research, IRB approval is crucial because most SoTL projects involve collecting data from students, including surveys, interviews, classroom observations, or analyzing student work, which qualifies as human subjects research. Obtaining IRB approval ensures that the research design minimizes risks to participants, secures informed consent, protects confidentiality, and complies with federal, institutional, and ethical standards. Even if the risks seem minimal, such as asking students about their learning experiences, approval is necessary because of the inherent power dynamics between faculty and students and the need to safeguard voluntary participation. Furthermore, IRB approval is often required to present or publish SoTL findings at conferences or in academic journals, making it an essential step for anyone intending to disseminate their work beyond the classroom.

10.5 SoTL Accelerator - Phase 3: Disseminate Findings

Drafting a SoTL-focused manuscript can be particularly difficult for engineering faculty because it requires a shift in thinking from technical research writing to educational scholarship, which often follows different conventions. Engineering faculty are typically trained to present data-driven, objective findings with a focus on technical rigor, whereas SoTL writing demands a balance among narrative, reflection, theoretical framing, and evidence of student learning. Many engineering faculty may also be less familiar with educational research methodologies, such as qualitative analysis or mixed methods, which are often expected in SoTL publications. Additionally, articulating the broader significance of a teaching intervention, beyond a single course or context, can be challenging, especially when engineering educators are more accustomed to writing technical papers with a narrow focus. Finding appropriate journals and navigating different peer-review expectations for SoTL work can add further complexity. Time constraints, limited institutional support for SoTL, and a lack of formal training in educational research writing also contribute to the difficulty, making it harder for engineering faculty to craft strong, publishable SoTL manuscripts.

A key resource in the dissemination process **SoTL Accelerator Tool #6: SoTL Template** (see https://www.sotlaccelerator.com/our-approach), shown in Fig. 10.2a, b, helps educators systematically investigate and reflect on their teaching practices and student learning experiences with the ultimate goal of improving educational outcomes. Using the SoTL template provides several key benefits: It offers a clear structure to guide the research process from formulating questions to reporting findings; ensures comprehensive coverage of important components such as literature review and methodology; promotes deep, reflective practice; supports effective documentation and dissemination of results; and facilitates collaboration among team members. By following a standardized framework (including main sections such as introduction, literature review, methods, results, discussion, and conclusion) educators can conduct more focused, valid, and publishable SoTL studies, thereby contributing meaningfully to the broader educational community.

10.6 Summary

Many new PhD graduates enter academia without sufficient preparation for teaching, largely because doctoral programs emphasize research development far more than instructional training. Additionally, participating in SoTL demands significant time, research skills, and institutional backing, resources that are often scarce in academic settings where conventional research achievements are prioritized. Moreover, it can be challenging to apply findings from classroom studies to broader educational contexts, especially when considering the varied backgrounds and needs of diverse student groups.

Title

- *The title summarizes the main idea or ideas of your study. A good title contains the fewest possible words that adequately describe the contents and/or purpose of your research paper.*

Abstract

- *The abstract provides an executive summary for the paper. The abstract typically summarizes the motivation, contribution, intervention and assessment, results, and practical takeaways.*

1. Introduction *[Entire Introduction should include at least 4-6 citations]*

1.1 Problem Identification

- *Problem Identification: This section should summarize the problem. At a minimum, the problem identification should include a sentence that directly states the problem (e.g., "The problem is..." or "The challenge is..."), should state what will happen if the problem doesn't get solved, and document the size of the problem. Statistics are beneficial.*

1.2 Study Overview

- *Study Overview: The Introduction should provide a summary of the solution proposed in response to the problem identified. This section should include a research question and/or research objectives. This section should provide an overview of the paper.*

2. Literature Review *[Each topic section should include at least 4-6 citations]*

2.1 [Topic - Expand on Problem]

- *Problem Identification: Expand on the problem. Depending upon the problem and paper focus, consider including the following information: explanation for project scope and focus area, definitions, assumptions, challenges, etc...*

2.2 [Topic - Expand on Current Approaches and Gaps]]

- *Current Approaches to the Problem: This section should explain the current approaches to solving the problem (ideally, a minimum of three current approaches to the problem).*
- *Gaps in Current Approaches: For each current approach mentioned, this section should provide a gap (e.g., reason why the approach doesn't completely solve the problem).*

2.3 [Topic - Expand on Proposed Solution (Intervention and/or Assessment)]

- *Expand on Study Overview to highlight the contribution. This can be done by explaining how the current study responds to the problem identified, how the solution is different from the current approaches, and how the current study overcomes the gaps presented.*
- *Add transition into Methods.*

3. Methods

- *The beginning of the methods section should provide an introduction to the study design via an overview of the participants, intervention, data collection instrument, and data analysis approach. Particular emphasis should be placed on order of events.*

Fig. 10.2 (**a**) SoTL Dissemination Template – Part 1. (**b**) SoTL Dissemination Template – Part 2

But a solution exists! The SoTL Accelerator, established in 2021 with funding from the Kern Family Foundation, is a free online resource available at https://www.sotlaccelerator.com/ that guides engineering educators through the SoTL process of curriculum development, assessment, and dissemination.

Through the SoTL Accelerator, engineering educators systematically study and refine their teaching practices, leading to more innovative, evidence-based instructional approaches such as project-based learning, active learning, and real-world problem integration. This continual improvement not only enhances student outcomes but also helps educators stay current with pedagogical research, accreditation expectations (such as engineering and technology accreditation organization ABET's emphasis on learning outcomes), and workforce demands.

Furthermore, engaging in SoTL fosters a culture of reflective practice, encourages interdisciplinary collaboration, and positions engineering faculty as leaders in educational innovation within their departments and professional societies. It ultimately strengthens both teaching effectiveness and career advancement by

3.1 Intervention

- *The intervention section should provide an over of the new curriculum change with enough information so that it can be replicated by another instructor. It may be helpful to leverage the Appendix to place assignment directions, example projects, etc...*

3.2 Participants

- *The participants section should provide a summary of the participants (e.g., makeup of entire class and makeup of those who completed the assessment) and context (e.g., course information).*

3.3 Data Collection Instrument(s)

- *The data collection section should provide a copy of the data collection instrument(s) and/or reference the data collection instrument(s) within the Appendix.*

3.4 Data Analysis Procedure(s)

- *The data analysis section should summarize the data analysis procedure(s) (e.g., thematic analysis).*

4. Results and Discussions

4.1 – 4.n [Summarize results in logical sequence]

- *The results should simply state the findings, interpret the findings (without bias, explanation, or justification), and be arranged in a logical sequence (e.g., by research question/objective, by instrument, by data analysis method, etc...). The results section should always be written in the past tense. It is NOT common to include citations in this section.*

4.4 Summary of Themes

- *The discussion should explain the findings, and provide plausible reasons for unexpected results.*
- *This section should provide a visual summary of the different themes.*

4.5 Lessons Learned

- *From an instructor perspective, What changes [if any] were made during the intervention? What will you do differently next time? What will you keep moving forward? What alternative teaching approaches [if any] are available?*

5. Conclusion

5.1 Response to Research Question/Objective and Summary of Main Takeaway

- *Respond to Research Question / Objective: The conclusion section should provide a summary response to the research question / objective.*
- *Emphasize Takeaway: The conclusion section should emphasize the main takeaway.*

5.2 Implications for Practitioners

- *Implications for Practitioners: The discussion section should describe *who* should do *what* differently as a result of this study. Be sure to weave in lessons learned.*

5.3 Limitations and Future Research

- *Limitations and Future Research: The conclusion section should highlight limitations of the study, and provide opportunities to overcome the limitations through future research.*

Fig. 10.2 (continued)

providing opportunities for scholarly publication, conference participation, and recognition of teaching excellence.

References

Bhanji, F., Gottesman, R., de Grave, W., Steinert, Y., & Winer, L. R. (2012). The retrospective pre–post: A practical method to evaluate learning from an educational program. *Academic Emergency Medicine, 19*(2), 189–194.

Boshier, R. (2009). Why is the scholarship of teaching and learning such a hard sell? *Higher Education Research & Development, 28*(1), 1–15.

Bosman, & Fernhaber. (2018). *Teaching the entrepreneurial mindset to engineers*. Springer International Publishing.

Buswell, N. T. (2021). The purpose of a PhD in engineering: Where does teaching fit in? *Studies in Engineering*. https://doi.org/10.21061/see.8

Coulter, S. E. (2012). Using the retrospective pretest to get usable, indirect evidence of student learning. *Assessment and Evaluation in Higher Education, 37*(3), 321–334.

Dickson, K. L., & Treml, M. M. (2013). Using assessment and SoTL to enhance student learning. *New Directions for Teaching and Learning, 2013*(136), 7–16.

Fukuzawa, S., Ashbourne, D., & Rawle, F. (2020). Overcoming challenges to impactful SoTL. In *Evidence-based faculty development through the scholarship of teaching and learning (SoTL)* (pp. 366–387). IGI Global.

Gansemer-Topf, A. M., McCloud, L. I., & Braxton, J. M. (2024). Defining the scholarship of teaching and learning (SoTL). *New Directions for Student Services, 2024*(185), 9–17. https://doi.org/10.1002/ss.20502

Garcia, O. J., & Kittur, J. (2024, 9–12 Dec. 2024). Factors influencing engineering doctoral students' perceptions on their readiness to teach engineering courses. *2024 IEEE international conference on teaching, assessment and learning for engineering (TALE)*.

Gilpin, L. S., & Liston, D. (2009). Transformative education in the scholarship of teaching and learning: An analysis of SoTL literature. *International Journal for the Scholarship of Teaching and Learning, 3*(2), 11.

Kennedy, D. R., Beckett, R. D., & O'donnell, L. A. (2020). Strategies, ideas, and lessons learned while engaging in SoTL without formal training. *American Journal of Pharmaceutical Education, 84*(1), 7702.

Kern, B., Mettetal, G., Dixson, M., & Morgan, R. K. (2015, 06/05). The role of SoTL in the academy: Upon the 25th anniversary of Boyer's scholarship reconsidered. *Journal of the Scholarship of Teaching and Learning, 15*(3), 1–14. https://doi.org/10.14434/josotl.v15i3.13623.

Linder, K. E., Elek, E. D., & Calderon, L. (2014). SoTL and the institutional review board: Considerations before navigating the application process for classroom research in higher education. *Journal of the Scholarship of Teaching and Learning, 14*(2), 1–14.

Rahimi, S., & Maloku-Morina, M. (2023). The usage of reflections on assessment process. *EDUCATION Journal of Educational Research, 5*(9-10), 201–206.

Rios, A. J., Plevris, V., Nogal, M., & Admiraal, W. (2025). Scholarship of teaching and learning in civil and structural engineering: A systematic literature review. *Journal of Civil Engineering Education, 151*(3), 03125001. https://doi.org/10.1061/JCEECD.EIENG-2103

Rüütmann, T., & Saar, M. (2017, 25–28 April 2017). Scholarly teaching and scholarship of teaching and learning in teaching engineering. *2017 IEEE global engineering education conference (EDUCON)*.

Wiggins, G. P., & McTighe, J. (2005). *Understanding by design*. ASCD Publications.

Wiliam, D. (2017). Assessment and learning: Some reflections. *Assessment in Education: Principles, Policy and Practice, 24*(3), 394–403.

Chapter 11
Example Curricula

Abstract Example curricula provide valuable insights into how other educators structure and sequence course content, which can inspire improvements in your own classroom. Such examples help identify best practices, innovative teaching methods, and effective integration of skills like design, communication, and teamwork. Additionally, reviewing examples can reveal gaps or redundancies in your current curriculum and guide evidence-based revisions aligned with accreditation standards or industry needs. From 2021 to 2023, the authors worked with more than 30 college-level engineering educators to pilot and validate the SoTL Accelerator (see Chap. 10). This chapter provides example curricula and links to the associated publications so readers can better understand the integrative and perspectives-taking approaches in action. Several examples can be found here: https://www.sotlaccelerator.com/example-papers.

Keywords Curriculum development · Learning objectives · Learning activities · Dissemination · Scholarship of teaching and learning (SoTL)

11.1 Introduction

> "I am not self-made. I am the product of many people who have given me opportunities and support to learn and grow."
>
> —*Michelle Obama, Lawyer and Former First Lady of the United States 2009-2017*

It is important for engineering faculty to have access to example curricula as they develop and refine their own course materials and programs. Reviewing well-designed curricula from peer institutions or colleagues provides valuable insight into effective content sequencing, assessment strategies, and alignment with accreditation standards and industry needs. **These examples serve as benchmarks and sources of inspiration, helping faculty identify gaps, adopt best practices, and incorporate innovative teaching approaches.** By examining how others structure learning experiences, faculty can make more informed decisions that enhance

L. Bosman et al., *Developing the Entrepreneurial Mindset in Engineering Students through a Perspective-Taking Approach*, Springer Texts in Education,
https://doi.org/10.1007/978-3-032-15123-0_11

course quality, support student success, and ensure programs remain current in an evolving engineering landscape.

11.2 Using a Da Vinci Biomimicry Approach to Explore Bio-Inspired Design in the Engineering Design Process

COURSE DETAILS:

Instructor: Prof. Randy Brooks	
Texas A&M University (College Station, TX, USA)	
https://www.linkedin.com/in/randybrookstamu86/	
Course: Engineering Mechanics	
Discipline: General Engineering	

SCHOLARLY DISSEMINATION:

Supplemental Teaching Resources	
https://engineeringunleashed.com/card/3168	

Peer-Reviewed Publication	
https://peer.asee.org/44566	

DESCRIPTION:

This project addresses student development of critical professional skills. Communication, teamwork, reflection, and the entrepreneurial mindset are all applied in the exploration of bio-inspired design and sustainability through an engineering design process (EDP) activity. Students experience new and innovative teaching and learning techniques while engaging with concepts that are new to them such as bio-inspired design and the entrepreneurial mindset. The final student artifact is an mp4 of the students presenting their EDP process and the resulting "big idea," which reflects the integration of the entrepreneurial mindset, bio-inspired design, and STEAM.

PERSPECTIVES-TAKING LEARNING OBJECTIVES:

By the end of this project, students will be able to:

- **Entrepreneurial mindset:** Critique product viability considerations for entrepreneurial advancement.
- **Bio-inspired design:** Apply biomimicry techniques to propose a solution focused on sustainability.
- **Arts Integration:** Communicate the key elements of your journey in a multimedia format art piece.

LEARNING ACTIVITY AND STUDENT DELIVERABLES:

This project, including project-related student deliverables, starts after students obtain an initial foundational knowledge in the course content area.

- **Week 4:** The photovoice assessment tool is new to the students, so this activity is introduced early in the semester in order to address challenges with the activity before being used in the bio-inspired design plan reflection in week 13.
- **Week 8:** (1) Students are grouped in teams of four per table. This will be the team that they go through the EDP with in weeks 8 and 10. (2) Students are introduced to the vision of bio-inspired design through some examples of biomimicry, and an introduction to Leonardo Da Vinci's work. (3) Students share with their table peers, then as a class. (4) The student exit ticket from each student is a sketch and description of their initial thought regarding a product to explore.
- **Week 10:** (1) Students dive into the bio-inspired EDP to include capturing their brainstorming ideas, then filtering those ideas to agree on a design to pursue. The week 10 teams are the same as the week 8 teams. (2) The student teams share digital snapshot of how far they got during the class time. This is a team-level exit ticket.
- **Week 11:** Students submit a digital multimedia artifact showing and describing their device to include the biomimicry relationship and a reflection on their process (brainstorming and filtering) to include entrepreneurial mindset elements.
- **Week 12:** Student products are posted in the digital discussion board for a gallery view/comment, and students are challenged to further explore both their current project and alternate applications of their product.
- **Week 13:** Administer a final photovoice assessment to encourage students to reflect on the impact of the process and target concepts.
- **Week 14:** Administer the open-ended question activity to spur student flection on the impact of the bio-inspired lesson on their thinking.

11.3 Encouraging Self-Directed Learning to Enhance Proficiency in Computer-Aided Design (CAD)

COURSE DETAILS:

Instructor: Dr. Thomas Omwando	
Simpson University (Redding, CA, USA)	
https://www.linkedin.com/in/thomas-omwando-ph-d-cem-141b9b40/	
Course: Computer-Aided Design (CAD) and Modeling	
Discipline: Engineering	

SCHOLARLY DISSEMINATION:

Supplemental Teaching Resources
https://engineeringunleashed.com/card/3632

Peer-Reviewed Publication
https://peer.asee.org/47604

DESCRIPTION:

Traditionally, introductory Computer-Aided Design (CAD) and Modeling courses teach students through drafting models and completing text-book-based exercises. While this method effectively builds foundational skills, it does not equip students with the entrepreneurial and professional competencies that are increasingly essential in today's engineering industry. To better prepare students for the demands of the job market, an integrated approach is introduced, incorporating the entrepreneurial mindset, bio-inspired design, and STEAM. This assignment is designed to foster creativity and problem-solving by encouraging students to identify issues in existing products and develop innovative solutions. The process involves product dissection, functional decomposition to analyze component relationships, and identification of gaps to improve existing or create entirely new designs. Students work in groups, which enhances their teamwork, communication, and critical-thinking skills. Their high levels of motivation, engagement, and investment in their projects significantly enriches their learning experience.

PERSPECTIVES-TAKING LEARNING OBJECTIVES:

By the end of this project, students will be able to:

- **Entrepreneurial mindset:** Identify and dissect an existing product/device to discover opportunities for improvement, or create a completely new design.
- **Bio-inspired design:** Analyze the functional components of the design, and identify and compare possible alternative environmentally sustainable materials or processes for manufacturing the proposed improvement or innovation.
- **STEAM:** Use computer aided design (CAD) tools to create infographics relevant to both the existing product and the improved/new design, and communicate the functional and aesthetic differences between the existing design and the improved/new design to various audiences.

LEARNING ACTIVITY AND STUDENT
- **DELIVERABLES:**

In addition to learning about computer-aided design (CAD) and finite element analysis (FEA) of devices/products in SolidWorks, this project aims to introduce students to the following:

- **Entrepreneurial mindset:** The first aim of this project is to identify and dissect an existing product/device to discover opportunities for improvement, or create a completely new design. The entrepreneurial mindset is a way of thinking that emphasizes creativity, innovation, risk-taking, and a focus on opportunity rather than on problems. Individuals with an entrepreneurial mindset tend to be proactive, persistent, and adaptable in the face of challenges. When engaging in a design thinking micro perspective exercise involving ideation, prototyping, and testing for products/devices, virtual simulation tools (e.g., CAD, FEA) can result in huge cost and time savings.

- **Bio-inspired design:** The second aim of this project is to analyze the functional components of the design, and identify and compare possible alternative environmentally sustainable materials or processes for manufacturing the proposed improvement or innovation. Bio-inspired design is an innovative approach that draws inspiration from nature to address engineering and design challenges. By studying natural systems, engineers and designers develop efficient and sustainable technologies through insights into materials, structures, and processes. This approach emphasizes systems thinking, requiring analysis across multiple levels to connect biological and engineering systems. Functional modeling in engineering mirrors the decomposition of biological systems, fostering a structured approach to innovation.
- **STEAM:** The third aim of this project is to use computer aided design (CAD) tools to create infographics relevant to both the existing product and the improved/new design, and communicate the functional and aesthetic differences between the designs to various audiences. There is global recognition of the growing importance of STEM education. Integrating STEAM into engineering incorporates art and design, helping students develop a stronger aesthetic sense and create more effective engineering solutions. It encourages interdisciplinary connections, broadening students' understanding of engineering challenges. By offering diverse tools and techniques, STEAM enhances problem-solving skills and introduces innovative approaches to engineering. The integration of the arts enables students to visualize STEM concepts creatively. Additionally, STEAM fosters digital literacy skills, such as coding and data analysis, which are essential in today's world. Embedding these skills in engineering education gives students a competitive advantage in the job market.

11.4 Nature-Inspired Manufacturing Podcast Creation to Bolster Communication Skills

COURSE DETAILS:

Instructor: Dr. Thomas Lucas	
Purdue University (West Lafayette, IN, USA)	
https://www.linkedin.com/in/tom-lucas-b827152b9/	
Course: Introduction to Industrial Controls	
Discipline: Engineering Technology	

SCHOLARLY DISSEMINATION:

Supplemental Teaching Resources	
https://engineeringunleashed.com/card/3148	

Peer-Reviewed Publication
https://peer.asee.org/47783

DESCRIPTION:

This project presents a semester-long asynchronous activity that asks students to research the origin and evolution of a bio-inspired manufacturing process. Students research the history of a process, compile and analyze their findings, and produce summaries of their findings in the form of a brief podcast episode intended for a general adult audience. This format requires the information from the early research steps to be understood and formatted into a storyboard, tuned through peer feedback, and produced into the episode.

PERSPECTIVES-TAKING LEARNING OBJECTIVES:

By the end of this project, students will be able to:

- **Entrepreneurial mindset:** Conduct market research to understand stakeholders and historical market trends associated with a manufacturing process.
- **Bio-inspired design:** Research the evolution of a bio-inspired manufacturing process.
- **STEAM:** Create a podcast to summarize the history and evoluation of a manufacturing process.

LEARNING ACTIVITY AND STUDENT DELIVERABLES:

The project can be assigned to students in a series of six steps. Students will need to:

1. Watch a video on an introduction to biomimicry. Perform research to find a manufacturing process that was inspired by a biological process. Summarize your findings in a short paragraph on a class discussion board. Review the other posts, and reply with your thoughts to two other posts with the least replies.
2. Watch a video on the introduction to skillset versus mindset. Focus on the modern applications of your process from the first step. Identify the companies, products, and industries that are affected by this process. Summarize your findings in a short paragraph on a class discussion board, review the other posts, and reply with your thoughts to two other posts with the least replies.
3. Watch a video on and introduction to storyboarding. Create a storyboard for your podcast episode.
4. Implement a tuning protocol to gather peer input, then storyboard the podcast episode again (or modify your original plan).
5. Record an audio file using the software of your choice. Consider segments such as an introduction, body, and closing remarks.
6. Produce and complete your final audio episode and submit the file to the instructor for upload to the podcast.

In this implementation, one podcast is created for the course and managed by the instructor. The students each produce episodes within the podcast.

11.5 Using Photovoice to Assess an REU Integrating Engineering, Energy, and Entrepreneurship

COURSE DETAILS:

Instructor: Dr. Bhavana Kotla	
The Ohio State University (Columbus, OH, USA)	
https://www.linkedin.com/in/bhavana-kotla-ph-d-984ba789/	
Course: Research Experience for Undergraduates (REU)	
Discipline: General Engineering	

SCHOLARLY DISSEMINATION:

Supplemental Teaching Resources	
https://engineeringunleashed.com/card/4703	

Peer-Reviewed Publication	
https://journals.sagepub.com/doi/abs/10.1177/15210251231192865	

DESCRIPTION:

The aim of this Research Experience for Undergraduates (REU) course is to provide underrepresented engineering and engineering technology students with a research experience that combines the best aspects of academic applied research, such as strong theory basis and rigorous scholarship, with essential business practices, including real-world customer discovery and the generation of sound business plans.

PERSPECTIVES-TAKING LEARNING OBJECTIVES:

By the end of this project, students will be able to:

- **Entrepreneurial mindset:** Apply human-centered design principles to evaluate needs from a customer discovery perspective.
- **Bio-inspired design:** Identify natural processes to develop solutions to pressing energy challenges.
- **STEAM:** Create a poster and flyer to visually communicate information relevant to the energy research project.

LEARNING ACTIVITY AND STUDENT DELIVERABLES:

Participants were expected to gain five core learning outcomes while working in a cohort of two to three students under the supervision of an advisor:

- Research skills development (e.g., data collection, data analysis, literature review, report writing, poster creation, etc.)
- Entrepreneurial mindset development (e.g., business model canvas, marketing analysis, customer discovery, patent search, etc.)

- Professional skills development (e.g., communication, collaboration, individual development, other transferable skills, etc.)
- Cohort/team relationship-building (e.g., team-building, networking, sense of belonging, etc.)
- Advisor relationship-building (e.g., "managing" your advisor, establishing trust, active listening, etc.)

During the 10-week summer experience, each week participants completed a weekly photovoice reflection. Photovoice uses photo (e.g., pictures and images) and voice (e.g., personal narrative) to reflect upon experiences. The photovoice reflection prompts focused, perceived changes in the development of an entrepreneurial mindset, research skills, professional skills, cohort/team relationship-building, and advisor relationship-building.

At the end of the experience, participants completed a final reflection assessment related to the five core learning objectives.

At the end of summer, all participants presented their research posters at the undergraduate research symposium.

11.6 Encouraging Self-Directed Learning to Enhance Proficiency in Computer-Aided Engineering (CAE) Design

COURSE DETAILS:

Instructor: Dr. Ashutosh Khandha	
University of Delaware (Newark, DE, USA)	
https://www.linkedin.com/in/ashutoshkhandha/	
Course: Medical Device Development Using Computer-Aided Engineering (CAE) and Modeling	
Discipline: Biomedical Engineering	

SCHOLARLY DISSEMINATION:

Supplemental Teaching Resources	
https://engineeringunleashed.com/card/3621	

Peer-Reviewed Publication	
https://2024bmesannual.eventscribe.net/fsPopup.asp?PresentationID=1504015&mode=presInfo	

DESCRIPTION:

A modern and iterative design approach is often lacking in Computer-Aided Engineering (CAE) projects in the Biomedical Engineering (BME) curriculum. Moreover, because BME curriculum typically conducts projects in team-based settings, there is the potential for individual students to remain underconfident in their abilities to use CAE tools, especially when faced with ill-defined problems in engineering. To that end, this assignment was introduced in a Medical Device Development course for juniors and seniors. The assignment is designed to help students gain experience and confidence in using CAE tools as an individual, open-ended, device design activity with minimal constraints.

PERSPECTIVES-TAKING LEARNING OBJECTIVES:

By the end of this project, students will be able to:

- **Entrepreneurial mindset:** Create a CAE design to optimize customer needs and manufacturability.
- **Bio-inspired design:** Apply systems thinking to better encourage ethical designs focused on sustainability while considering unintended consequences.
- **STEAM:** Critique art sculptures and/or architecture to gain inspiration for the CAE strut design.

LEARNING ACTIVITY AND STUDENT DELIVERABLES:

In addition to learning about computer-aided engineering (CAE) design and finite element analysis (FEA) of medical devices in SolidWorks, this project aims to introduce students to the following:

- **Entrepreneurial mindset:** Simply defined, an entrepreneurial mindset is the inclination to discover, evaluate and exploit opportunities. People with an entrepreneurial mindset think strategically about innovation and continuous improvement, regardless of whether they work in a large company or a start-up. When engaging in a design thinking micro perspective exercise involving ideation, prototyping, and testing for medical devices, virtual simulation tools (e.g., CAD, FEA) can result in huge cost and time savings. The first aim of this project is to practice CAE design and FEA for a specific problem by considering business and value perspectives.
- **Bio-inspired design:** By harnessing the natural design rules of the very biology that is life on the planet, and by translating solutions from one area to another, new pathways can be found to address complex challenges. The second aim of this project is to practice CAE design and FEA for the specific problem (defined in the assignment), by consciously being aware of bio-inspired design and biomimicry.
- **STEAM:** There is global recognition of the growing importance of STEM education. Merging the arts with STEM for design, invention, and innovation activities can lead to enhanced creativity and problem-solving skills. As an example, engaging with visual arts such as sculptures and architecture have the potential to conceive unique, aesthetic and optimized solutions for orthoses and prostheses. The third aim of this project is to practice CAE design and FEA for the specific problem (defined in the assignment), by drawing inspiration from the visual arts.

11.7 Engage Engineering Students in Teaching Linear Elasticity Through Entrepreneurially Minded, Bio-Inspired Projects

COURSE DETAILS:

Instructor: Dr. Jeff Ma	
Saint Louis University (St. Louis, MO, USA)	
https://www.slu.edu/science-and-engineering/academics/aerospace-and-mechanical-engineering/faculty/ma-jeff.php	
Course: Computer-Aided Engineering	
Discipline: Mechanical and Aerospace Engineering	

SCHOLARLY DISSEMINATION:

Supplemental Teaching Resources	
https://engineeringunleashed.com/card/3422	

Peer-Reviewed Publication	
https://peer.asee.org/43388	

DESCRIPTION:

Some construction biomaterials have superior strength and/or hardness. For example, biotech concrete incorporates ureolytic bacteria and tough materials such as spider silk and nacre shells to improve the strength. In this module, students learn the fundamentals of elasticity of materials with respect to analytically and numerically analyzing mechanical components. Students investigate how animals (such as spiders) can generate these superior construction materials and how people can take advantage of them in the construction of buildings.

PERSPECTIVES-TAKING LEARNING OBJECTIVES:

By the end of this project, students will be able to:

- **Entrepreneurial mindset:** Evaluate a structure both analytically, using equations, and numerically, using Abaqus modeling software, to predict the soundness of this structure for commercial purposes.
- **Bio-inspired design:** Describe current applications of bio-inspired architecture materials and the challenges of these materials.
- **STEAM:** Incorporate functional aesthetics into the design models.

LEARNING ACTIVITY AND STUDENT DELIVERABLES:

The first module in this Computer-Aided Engineering (CAE) course, summarized here, is linear elasticity of engineering materials.

Module Milestones:

- Part 1: Students are asked to investigate several construction biomaterials that have superior strength and/or hardness.

- Part 2: Students are asked to investigate the possibility of applying these biomaterials into constructions that people can use (e.g. buildings, sculptures and statues, monuments, etc.) for superior durability without plastic deformation.
- Part 3: The students are asked to design/analyze constructions built from the investigated biomaterials via analytical tools (equations) and a numerical tool (ABAQUS FEM software). The students will also predict the artistic values of the constructions.

Module Learning Activities:

- Learn stress and strain concepts;
- Learn equilibrium equations in terms of stress components;
- Learn kinematic equations;
- Learn constitutive equations;
- Learn compatibility equations; and
- Learn Abaqus modeling of elastic deformation of structures.

Module Assessment of Learning:

- Report and presentation about designed construction/structure made of biomaterials using equations and/or using Abaqus, marker analysis, and cost analysis of the constructure/structure.

11.8 Music of the Heart: Identification and Inspiration from Musical Elements in Normal and Diseased Heart Sounds

COURSE DETAILS:

Instructor: Dr. Sabia Abidi	
Rice University (Manvel, TX, USA)	
https://www.linkedin.com/in/sabia-abidi-a5857116/	
Course: Fundamentals of Systems Physiology	
Discipline: Bioengineering	

SCHOLARLY DISSEMINATION:

Supplemental Teaching Resources	
https://engineeringunleashed.com/card/3195	

Peer-Reviewed Publication	
https://peer.asee.org/43675	

DESCRIPTION:

Students focus on evaluating normal and diseased heart sounds from a macro perspective. There is also an opportunity for students to define normal versus diseased sounds and ideate/create a musical composition that reflects those musical features. The production of a musical piece inspired by heart sounds incorporates the STEAM aspect. The ability to diagnose the heart based on sounds fits well into the biomedicine and health outcome via bio-inspired integration. Students can apply this project to other relevant physiological sounds (e.g., lungs, abdominal cavity, baby) and interpret the data. Because the activity is conducted in groups, students have the ability to discuss and conduct the exercise in a safe environment without feeling overly taxed. The exercise gives students an opportunity to be creative by translating physiologically created sounds to instrument of their choice. Interpretation of heart sounds is a relevant aspect in our cardiovascular unit but is unfortunately not studied due to time constraints. This exercise gives students an opportunity to explore the heart sounds in a fun, creative way.

PERSPECTIVES-TAKING LEARNING OBJECTIVES:

By the end of this project, students will be able to:

- **Entrepreneurial mindset:** Apply engineering principles to detect and design a training model for physicians.
- **Bio-inspired design:** Promote pattern recognition in varying contexts, ability to diagnose pathology based on sound information.
- **STEAM:** Identify differentiating musical elements in normal versus diseased hearts, and create an expression of the normal or diseased heart sound using instruments available at home or via publicly available software.

LEARNING ACTIVITY AND STUDENT DELIVERABLES:

Student teams ($n = 4$) have one month to complete this assignment. Simultaneous early class lectures aid in physiological context. Overall, the exercise is designed to encourage pattern recognition in disparate contexts and diagnosis of pathology based on sound information.

- Part 1: Students articulate the differences between a normal and diseased heart sound.
- Part 2: Students connect differences in heart sounds to physiological causes.
- Part 3: Students identify differentiating musical elements in normal vs diseased heart.
- Part 4: Students create an expression of the normal or diseased heart sounds using instruments available at home or via publicly available software.
- Part 5: Students apply engineering principles to detect and design a training model for physicians.

11.9 Inspiring the Entrepreneurial Mindset in Engineering Students Using Project-Based Learning Approach in IoT Systems Design Class

COURSE DETAILS:

Instructor: Dr. Hadil Mustafa	
California State University – Chico (Chico, CA, USA)	
https://www.linkedin.com/in/hadil-mustafa-0a65202/	
Course: IoT Systems Design	
Discipline: Electrical and Computer Engineering	

SCHOLARLY DISSEMINATION:

Supplemental Teaching Resources	
https://engineeringunleashed.com/card/3489	

Peer-Reviewed Publication	
https://peer.asee.org/43979	

DESCRIPTION:

This six-week project starts with students searching and evaluating available IoT systems in the market targeting house or farm animals' security. This approach allows students to identify the need and marketability of their proposed designs and establish a better understanding of the system requirements. All proposed designs must incorporate one bio-inspired aspect and satisfy all IoT systems requirements, including networking, user-friendly interface, sensing, and actuating. The final product is evaluated on originality, creativity, system integration, functionality, and ease of use.

PERSPECTIVES-TAKING LEARNING OBJECTIVES:

By the end of this project, students will be able to:

- **Entrepreneurial mindset:** Use public application programming interface and software development kit to design an embedded system with networking capabilities.
- **Bio-inspired design:** Evaluate current commercially available bio-inspired IoT systems.
- **STEAM:** Apply media artworks to reflect purpose, function, and audience in IoT systems.

LEARNING ACTIVITY AND STUDENT DELIVERABLES:

- **Milestone 1: Research and analysis.** The objective of this milestone is to introduce students to bio-based IoT systems in the market. Students will form groups to search, select, and analyze one IoT system targeting animals' security or safety. All analyses should contain a description of the system's interfaces and classifications of inputs/outputs based on their mode of interaction.

- **Milestone 2: Project proposal.** The objective of this milestone is to promote and instill an entrepreneurial mindset through the planning and development of the system design. All groups should select a design that can fit in one of the bio-inspired fields: sustainability, biomedicine, or security. Students should interview at least two potential clients to help them analyze the market availability and field feasibility of their proposed design.
- **Milestone 3: Design prototype.** To execute this milestone, students should finish the following tasks: (1) complete the design prototype, (2) identify hardware and software components, (3) identify hardware and software interfaces, and (4) develop the system's design algorithm.
- **Milestone 4: Complete design and presentation.** To complete this milestone, each group should prepare a 15-minute presentation of their working design and submit a photovoice reflection assessment report. The completed design will be graded based on functionality, originality, number of hardware interfaces, quality of the design, and quality of the software. All students will submit a peer-evaluation report of their team members.

11.10 Integrating Entrepreneurship and STEAM with Green (Bio-Inspired) Roof Design

COURSE DETAILS:

Instructor: Dr. Nadia Al-Aubaidy	
Norwich University (Northfield, VT, USA)	
https://www.linkedin.com/in/dr-nadia-a-al-aubaidy-542375296/	
Course: Capstone Design	
Discipline: Civil Engineering	

SCHOLARLY DISSEMINATION:

Supplemental Teaching Resources	
https://engineeringunleashed.com/card/3633	

Peer-Reviewed Publication	
https://peer.asee.org/46544	

DESCRIPTION:

This project requires students to conduct research on a real-world residential construction project from concept to completion. Students explore bio-inspired innovative design ideas that can be integrated into building projects in response to human needs and environmental protection. Students evaluate sustainable features' aesthetic and economic viability toward making an informed decision that creates value for individuals, groups, and communities. The project's micro (the individual level) is directed at preparing a future independent researcher. The project's mezzo (the group level) enhances teamwork and collaboration across disciplines and offers a wide range of topics for discussion and research.

PERSPECTIVES-TAKING LEARNING OBJECTIVES:

By the end of this project, students will be able to:

- **Entrepreneurial mindset:** Explain how to conduct a business viability study, including market analysis and SWOT (strengths, weaknesses, opportunities, and threats) analysis.
- **Bio-inspired design:** Understand the impacts of design decisions from an environmental and natural world perspective.
- **STEAM:** Identify opportunities to integrate functional aesthetics into building designs.

LEARNING ACTIVITY AND STUDENT DELIVERABLES:

This assignment contributes to preparing today's students to be future green entrepreneurs and T-shaped engineers. The assignment is divided into four-week deliverables. Before introducing the final project and presentation, the knowledge and topics required to complete this assignment are scheduled to be covered in lectures. Green building concepts are introduced, and green roof systems are explained to students. Bio-inspired design is also demonstrated and supported by case studies connecting design with nature. Moreover, the topics of entrepreneurship, market analysis, and SWOT are presented, defined, and explained.

- **Week 1 (Research):** Students research suitable building locations in the US, green roof materials, vegetation types, bio-friendly irrigation methods, maintenance protocols, solar panel integration, and construction methods.
- **Week 2 (Market analysis and economic feasibility):** Students conduct market analysis, calculate cost estimates, and draft an economic feasibility study.
- **Week 3 (3D building information modeling module):** Students design 3D green roofs using BIM software, conduct a constructability review, and obtain feedback from potential stakeholders on the green roof design.
- **Week 4 (Final project and presentation):** Students identify areas for design improvement, submit a final report, and conduct a final PowerPoint presentation.

11.11 Contextual Learning of Surface Phenomena in Manufacturing Processes Through Entrepreneurially Minded, Bio-Inspired Design

COURSE DETAILS:

Instructor: Dr. Beshoy Morkos	
University of Georgia – Athens (Athens, GA, USA)	
https://www.linkedin.com/in/beshoy-morkos-24b50013/	
Course: Manufacturing Processes	
Discipline: Mechanical Engineering	

SCHOLARLY DISSEMINATION:

Supplemental Teaching Resources	
https://engineeringunleashed.com/card/3420	

Peer-Reviewed Publication	
https://peer.asee.org/44653	

DESCRIPTION:

This project introduces students to the use of a manufacturing survey instrument to understand various manufacturing processes and phenomena through personal exploration and inquiry. Each week, students complete a manufacturing survey that reflects what they learned in class that week. In doing so, they explore any product they see in use and determine if that product could have been manufactured using the process we learned about that week in class. Students survey various components and make a determination about their manufacturing process. The survey requires justification of the manufacturing process (e.g., how do you know it was made using die casting?), explanation of other manufacturing processes that could be used (e.g., forging, rolling, etc.), and identifying the parameters that control which process should be used.

PERSPECTIVES-TAKING LEARNING OBJECTIVES:

By the end of this project, students will be able to:

- **Entrepreneurial mindset:** Describe the effect of product design on manufacturability.
- **Bio-inspired design:** Identify products manufactured through a process motivated by biomimicry.
- **STEAM:** Explain the role of aesthetics in process functionality and design for manufacturability.

LEARNING ACTIVITY AND STUDENT DELIVERABLES:

- **Step 1: Identify a product.** Students are required to identify a product that was manufactured using a bio-inspired process. This allows them to both develop formal manufacturing knowledge and skills related to their hobbies or interests and cultivate an entrepreneurial mindset by considering different manufacturing processes that could improve the economic or social value of a product. The assignment also incorporates elements of aesthetics, which encourages students to think beyond product functionality and consider design aspects. This combination of biomimicry, entrepreneurial mindset, and art adds a unique perspective to traditional manufacturing processes education.

- **Step 2: Assess the product.** The manufacturing assessment survey asks students to: (1) summarize the content learned in classroom that week, (2) identify a product that they believe was made using that process (and attach a picture of the product), (3) provide physical justification/evidence for why the product was made in the manner it was made, and (4) describe and justify other processes that could be used to make the same product.
- **Example:** If students learned about biomachining this week, they would be expected to find products that exhibit properties similar to biomachining such as erosion, corrosion, and pitting.

11.12 Using Art, Entrepreneurship, and Bio-Inspired Design to Select Materials for a Simplified Bicycle Design

COURSE DETAILS:

Instructor: Dr. Gordon Krauss	
Harvey Mudd College (Claremount, CA, USA)	
https://www.linkedin.com/in/gordon-krauss-36b5632/	
Course: Introduction to Design	
Discipline: Materials Engineering	

SCHOLARLY DISSEMINATION:

Supplemental Teaching Resources	
https://engineeringunleashed.com/card/3199	

Peer-Reviewed Publication	
https://peer.asee.org/46459	

DESCRIPTION:

This is a design exercise based on material selection for a simplified bicycle design. The overall goal is to create a bicycle that better serves a (fictitious) defined customer segment whose financial, performance, and environmental consideration needs are underserved by existing market offerings. The performance of various existing bicycles is explored to motivate the exercise.

The exercise is appropriate for students in an Introduction to Materials Engineering course who have an understanding of stress and strain, cantilever beam deflection, fatigue, creep and relaxation, and impact (simply modeled as a Charpy-type test). These conditions are modeled in simplified versions of bicycle design.

The fundamental purpose of the exercise is for students to complete an engineering analysis of materials under different simplified loading conditions and select the best material from a provided list. Their decision should be based on functional performance, design objectives, and constraints, and may include conflicting objectives (e.g., be lightweight and be low cost). The intent of such conflicts is to require students to make a decision and calculate their assumptions with respect to the financial value of weight reduction.

PERSPECTIVES-TAKING LEARNING OBJECTIVES:

By the end of this project, students will be able to:

- **Entrepreneurial mindset:** Justify design selection by taking into consideration business viability, system constraints, and economic sustainability.
- **Bio-inspired design:** Describe design decisions in terms of quantified tradeoffs between objectives, including environmental sustainability.
- **STEAM:** Explain design decisions in terms of aesthetics and visual appeal.

LEARNING ACTIVITY AND STUDENT DELIVERABLES:

- **Part 1: Design activity.** Students are provided with the description of an engineering system; in this case, a simplified bicycle. The design objectives, functions, and constraints are listed for the overall system as well as individual components of interest (per Dym's design definition method). A simplified description of a component is used to conduct an engineering analysis of the component to select the best material and geometry combination to meet the objectives, functions, and constraints. Ideally, this is a nontrivial task, as different choices will optimize different objectives. This will require knowledge of the material properties and mechanics of the system (Young's modulus, yield stress, ultimate tensile stress, fatigue S-N curves, thermal creep and relaxation, etc.). Note that significant simplification is possible (such as ignoring shear).
- **Part 2: Bio-inspired design modification.** Students are asked to apply biomimicry concepts to improve the bicycle performance with respect to weight, cost, environmental impact, or durability.
- **Part 3: Infographic development.** Students create an infographic demonstrating the value of this modification that includes justification for the specific improvement, required engineering analysis approach, anticipated costs associated with the modification, and a sketch of the design improvement.

11.13 Cars of the Next Century: Using Photovoice as a High-Impact Practice in Engineering Technology

COURSE DETAILS:

Instructor: Dr. Khalid Tantawi	
The University of Tennessee at Chattanooga (Chattanooga, TN, USA)	
https://www.linkedin.com/in/khalidtantawi/	
Course: Principles of Mechanical Systems	
Discipline: Mechanical Engineering Technology	

SCHOLARLY DISSEMINATION:

Supplemental Teaching Resources	
https://engineeringunleashed.com/card/3429	

Peer-Reviewed Publication	
https://peer.asee.org/43652	

DESCRIPTION:

Scientists estimate that the global population will reach 11.2 billion by the year 2100, and that some super-large cities will have populations of as many as 90 million inhabitants. In the US, for example, more than 30 million people are expected to live in New York City by that same year. However, the widths of the streets of the major cities will not increase proportionally with the population. Therefore, because traffic jams will constitute a major challenge, experts believe that transportation will be the primary concern within the cities during the next century. Furthermore, the median life expectancy in the US is expected to reach 100 years or more. As a result, reducing traffic jams and heat emissions, improving air quality, and maintaining passenger comfort will be some of the main points to focus on in the design. In response to this challenge, students are tasked with designing cars that would be able to drive in twenty-second-century city streets. The students use artistic tools to display the design and present on the details of the car designs. Designs include modular vehicles, super-enhanced efficiency combustion engines, and hydrogen cells.

PERSPECTIVES-TAKING LEARNING OBJECTIVES:

By the end of this project, students will be able to:

- **Entrepreneurial mindset:** Describe the trends in technology for improving human lives based on publicly available scientific data and business viability.

- **Bio-inspired design:** Apply principles of biomimicry to identify environmental design opportunities.
- **STEAM:** Create a visual presentation to share the computer-aided design (CAD) drawings and car design.

LEARNING ACTIVITY AND STUDENT DELIVERABLES:

- **Research:** Students will explore and research mechanical technologies that have a potential to be used in the vehicles of the next century to solve the problem of overcrowding in super-large city road networks.
- **Mechanical design constraints:** To meet the course learning outcomes, technical constraints include required mechanical systems that have the potential to reduce the problems of traffic jams and air quality in a design of the vehicle of the next century.
- **Bio-inspired design constraints:** The conceptual design should consider applications of biomimicry to identify bio-inspired design opportunities.
- **Aesthetic design constraints:** The design should consider aesthetics expressed through an artistic drawing and/or animations or other means to convey the idea to others.
- **Presentation:** Students will present the research via a PowerPoint presentation and computer-aided design (CAD) drawings.

11.14 A Perspectives-Taking Approach to Biomedical Engineering Prosthetic Design

COURSE DETAILS:

Instructor: Dr. Adel Alhalawani	
Rose-Hulman Institute of Technology (Terre Haute, IN, USA)	
https://www.linkedin.com/in/adelalhalawani/	
Course: Mechanics of Materials	
Discipline: Biomedical Engineering	

SCHOLARLY DISSEMINATION:

Supplemental Teaching Resources	
https://engineeringunleashed.com/card/3574	

Peer-Reviewed Publication	
https://peer.asee.org/46472	

DESCRIPTION:

Many engineering students are underprepared for the workforce due to a lack of real-world application in the college curriculum. Consequently, the transition between college and the first job can be rough for many graduates, which causes some to seek jobs outside of the engineering field altogether. To combat this issue, a multidisciplinary project experience was implemented into the standard engineering curriculum that involves skills used in business, art, environmental science, and engineering. Specifically, students work on a prosthetic design.

PERSPECTIVES-TAKING LEARNING OBJECTIVES:
By the end of this project, students will be able to:

- **Entrepreneurial mindset:** Conduct a mechanical analysis on a below-elbow prosthetic tube to compare the properties of natural and human-made materials, revealing their comparative efficiencies from a business viability perspective.
- **Bio-inspired design:** Identify ways in which basic structural building blocks of the natural world (minerals, polysaccharides, and proteins) combine to create diverse biomaterials.
- **STEAM:** Describe how aesthetics can be integrated into engineering design projects.

LEARNING ACTIVITY AND STUDENT DELIVERABLES:

- **Select a pose:** Watch a kid's breakdance battle video and extract a stance/move in which the kid would apply some sort of a torsional load if he/she was wearing the prosthetic. Keep analysis simple! Consider only one torsional load. Ignore axial and bending loads. Consider the tubular section of the prosthetic only without including the hand attachment or the tube cap. Hint: You can assume a load is applied on the hand at a certain distance from the tubular section of the prosthesis.
- **Bio-inspired design:** Watch a video on the introduction to biomimicry. Choose at least two materials (one natural and one human-made) for the tubular section of your prosthetic. You should specify a safe design for each material, then report which design is lighter. You will need to support your hypothesis and justify the material selection.
- **Structural design:** The tubular section of your prosthetic must be able to withstand the force applied during the activity. Remember that *torque* is a measure of the force that can cause an object to rotate about an axis, so you need a linear force and a distance vector.
- **Entrepreneurial mindset:** Watch a video on the entrepreneurial mindset and reflect on its importance for you as a designer and engineer. The material should provide the lightest weight possible. This is based on the feedback you received from your customer. In other words, you should aim for the most valuable design.
- **Dance-inspired STEAM design:** Suggest an artistic addition to make your prosthetic aesthetically pleasing. You should communicate your artistic addition by drawing it. Identify something in the dance and respond to it in your design. Analyze the artistic expression of a dance. Discuss insights using evaluative criteria and dance terminology.

11.15 Entrepreneurially Minded Design of Machinery Through a Bio-Inspired Design Project with Aesthetic Objectives

COURSE DETAILS:

Instructor: Dr. Soheil Fatehiboroujeni	
Colorado State University (Fort Collins, CO, USA)	
https://www.linkedin.com/in/sfatehiboroujeni/	
Course: Dynamics of Machines	
Discipline: Mechanical Engineering	

SCHOLARLY DISSEMINATION:

Supplemental Teaching Resources	
https://engineeringunleashed.com/card/3427	

Peer-Reviewed Publication	
https://peer.asee.org/43040	

DESCRIPTION:

One of the key elements of the design of machinery is full kinematic and full dynamic analysis of mechanisms and their elements. This involves analytic modeling with the use of computational as well as CAD software. However, in practice, all such engineering analyses become the basis of design decisions in light of wide-ranging objectives that include human or user experience, health and safety, or aesthetic and marketability. This project creates an opportunity for engineering students to gain firsthand experience in how math modeling, simulations, and synthesis of machinery is integrated with (1) entrepreneurial aspects of design thinking, (2) health and bio-inspired dimensions of human-machine interface, and (3) aesthetic and artistic facets of their design as a whole.

PERSPECTIVES-TAKING LEARNING OBJECTIVES:

By the end of this project, students will be able to:

- **Entrepreneurial mindset:** Explore a variety of applications and products for which linkages and mechanisms are used to generate desirable motion, function, or output path.
- **Bio-inspired design:** Identify bio-inspired processes or health and biomedicine applications that pertain to dynamics of machines.
- **STEAM:** Describe aesthetic factors of mechanical systems.

LEARNING ACTIVITY AND STUDENT DELIVERABLES:

Students spend about 4 weeks on this activity. The key deliverable is a written technical report with given instructions on the form and required content.

I. **Introduction and need statement:** Introduce the purpose of your project. It is important to describe for what need or requirements you are designing your linkage. To establish your need, find something in your everyday life (or someone else's) that could be improved using a linkage mechanism. While this freedom may be challenging at first, it is typical of most engineering projects.
II. **Design thinking:** Engineering projects are embedded in markets and entrepreneurial contexts. You should implement the design thinking cycle resulting in at least two alternatives based on research of current state-of-the-art, available products, patents, etc.
III. **Position synthesis of the first prototype:** For your position synthesis, you must provide at least two prototypes of a two-dimensional, four-bar linkage to meet your need. To generate the two designs, you must use the three-position synthesis technique, whereby you must first determine the three most critical positions for your linkage.
IV. **Photovoice reflection:** This serves as your conclusion section to reflect on your own activities and their outcomes.
V. **Appendix:** You should include any calculations or designs that are too burdensome for the general body of the report but that you think will provide good insight into your design process.

11.16 Biologically Inspired Pandemic Preparedness Project to Instill the Entrepreneurial Mindset

COURSE DETAILS:

Instructor: Dr. Sujata Bhatia	
Harvard University (Cambridge, MA, USA)	
https://www.hks.harvard.edu/about/sujata-k-bhatia	
Course: Sophomore Engineering Clinic	
Discipline: General Engineering	

SCHOLARLY DISSEMINATION:

Supplemental Teaching Resources	
https://engineeringunleashed.com/card/3542	

Peer-Reviewed Publication	
https://ieeexplore.ieee.org/document/6207095	

DESCRIPTION:

Undergraduate engineering degree programs struggle to attract and retain a diverse student population. One potential strategy for addressing this problem is to empower engineering students to address pressing societal needs. The goal of this work is to implement a bio-inspired design project, focused on pandemic preparedness and the entrepreneurial mindset, in a sophomore engineering clinic at a large mid-Atlantic public university.

PERSPECTIVES-TAKING LEARNING OBJECTIVES:

By the end of this project, students will be able to:

- **Entrepreneurial mindset:** Explore public health related entrepreneurial designs to prepare for a future pandemic.
- **Bio-inspired design:** Analyze opportunities to take inspiration from biology to prepare for a future pandemic.
- **STEAM:** Create infographics to visually communicate information relevant to biomedical imaging.

LEARNING ACTIVITY AND STUDENT DELIVERABLES:

- The course incorporates a series of lectures on entrepreneurship, including an overview of disruptive innovation, the innovator's dilemma, and resistance to innovation.
- The course then incorporates a series of lectures on bio-inspired technologies, including brain-computer interfaces, gene editing, and medical diagnostics. Students work in multidisciplinary groups to apply information from lectures and develop entrepreneurial designs to prepare for a future pandemic.
- The students' creativity and originality, along with their collaborative spirit, combine to generate designs that would effectively serve society. Also, the project focuses directly on public health, thereby empowering engineering students to consider service to society.
- Data are collected via student self-assessment of learning using a standardized qualitative instrument. Specifically, students complete a metacognitive reflection assessment, which asks participants to respond to photovoice reflection prompts and open-ended questions.

11.17 Infographic Design for Artificial Intelligence and Biomedical Imaging

COURSE DETAILS:

Instructor: Dr. Lisa Bosman	
Purdue University (West Lafayette, IN, USA)	
https://www.linkedin.com/in/lisabosman/	
Course: Introduction to Engineering	
Discipline: General Engineering	

SCHOLARLY DISSEMINATION:

Supplemental Teaching Resources	
https://engineeringunleashed.com/card/3022	

Peer-Reviewed Publication	
https://doi.org/10.1515/edu-2022-0187	

DESCRIPTION:

This infographic development project requires students to visually communicate a variety of aspects at the intersection of artificial intelligence and biomedical imaging. This lesson plan can be completed as a one-time assignment or required multiple times throughout the semester to provide students practice with communicating through infographics. This project leverages Canva, a freely available online graphic development platform. Students create a comparison, statistical, timeline, process, and bio-inspired design infographic.

PERSPECTIVES-TAKING LEARNING OBJECTIVES:

By the end of this project, students will be able to:

- **Entrepreneurial mindset:** Identify, compare, and contrast ethical concerns associated with diagnostic biomedical imaging versus the traditional diagnostic approaches.
- **Bio-inspired design:** Identify functional bridges connecting biology to biomedical imaging.
- **STEAM:** Create infographics to visually communicate information relevant to biomedical imaging.

LEARNING ACTIVITY AND STUDENT DELIVERABLES:

- **Timeline infographic:** Create an infographic highlighting how diagnostic biomedical imaging has changed over time. The graphic should identify when different types of diagnostic biomedical imaging has been introduced to the public.
- **Process infographic:** Create an infographic to showcase the ethical issues associated with artificial intelligence and biomedical imaging (e.g., potential areas of error and consequences of diagnosing false positives and false negatives).
- **Statistical infographic:** Create an infographic that integrates statistical data from multiple sources to tell a story explaining the customer type(s) and other

market information (e.g., size, location, need) related to a diagnostic biomedical imaging technology.
- **Comparison infographic:** Create an infographic that provides an unbiased ethical comparison of diagnostic biomedical imaging versus the traditional diagnostic approaches (e.g., specific use of CT, MRI, PET, or ultrasound).
- **Bio-inspired design infographic:** Create an infographic to show how diagnostic biomedical imaging can take inspiration from biology.

11.18 Summary

In summary, reviewing example curricula can help educators refine their own teaching by showcasing how others organize content, apply effective practices, and incorporate essential skills. By comparing approaches, instructors can identify opportunities for improvement and alignment with professional standards. From 2021 to 2023, more than 30 engineering faculty collaborated to test and refine the SoTL Accelerator, which supports this kind of curriculum development. This chapter includes sample curricula and related publications to illustrate integrative and perspectives-based teaching strategies. Examples are available at sotlaccelerator.com/example-papers.

Chapter 12
Additional Resources

Abstract Extending this practical guidance, additional resources are offered to help educators further implement, adapt, and sustain these approaches in their own teaching and curriculum development. Professors seeking to incorporate the entrepreneurial mindset into their teaching can leverage a variety of valuable resources, including conferences, academic journals, professional networks, and online platforms. This chapter explores key opportunities for faculty development and curriculum enhancement by highlighting major conferences that provide insights into best practices, emerging research, and innovative teaching methodologies. In addition, the chapter examines influential academic journals that publish cutting-edge studies on entrepreneurial education. The chapter also showcases online resources that offer open-access materials, case studies, and tools for classroom implementation. Furthermore, the chapter discusses professional organizations, workshops, and collaborative platforms that provide networking opportunities and ongoing support for faculty. By utilizing these diverse resources, educators can effectively foster creativity, problem-solving skills, and an entrepreneurial mindset in their students, preparing them for success in an ever-evolving professional landscape.

Keywords Scholarship of teaching and learning (SoTL) · Conferences · Journals · Dissemination · Online platforms · Networking

12.1 Introduction

> "When we know better, we do better."
> —*Maya Angelou, American memoirist, poet, and civil rights activist*

Professional development for engineering faculty is essential to enhancing both teaching effectiveness and student learning outcomes. **As engineering fields rapidly evolve with new technologies and methodologies, faculty must stay current to deliver relevant and impactful instruction.** Engaging in professional development—through conferences, journal articles, professional networks, and online platforms—allows

L. Bosman et al., *Developing the Entrepreneurial Mindset in Engineering Students through a Perspective-Taking Approach*, Springer Texts in Education,
https://doi.org/10.1007/978-3-032-15123-0_12

faculty to exchange ideas, adopt innovative teaching practices, and remain informed about the latest research and industry trends. These resources support continuous growth, foster interdisciplinary collaboration, and enable faculty to design more engaging, evidence-based learning experiences that better prepare students for the demands of modern engineering careers.

12.2 Conferences

Attending conferences is a vital component of faculty development, particularly for educators seeking to integrate the entrepreneurial mindset into the engineering classroom. These events provide a dynamic platform for professors to engage with thought leaders, share best practices, and explore innovative teaching strategies that cultivate creativity, critical thinking, and problem-solving skills in students. Conferences foster collaboration by connecting faculty with like-minded educators, industry experts, and researchers who are shaping the future of entrepreneurship in engineering education. They offer hands-on workshops, keynote presentations, and panel discussions that equip instructors with practical tools and fresh perspectives. By participating in these gatherings, faculty members remain informed about emerging trends, gain inspiration from real-world case studies, and develop actionable strategies to enhance student learning. Ultimately, **conferences serve as a catalyst for continuous improvement**, ensuring that educators can effectively prepare students for the complex, ever-evolving challenges of the engineering profession (Table 12.1).

Table 12.1 Example conferences

Conference	URL
KEEN National Conference	https://engineeringunleashed.com/keen-national-conference
VentureWell OPEN Annual Conference	https://venturewell.org/open/
Babson College Entrepreneurship Research Program & Conference (BCERC)	https://www.babson.edu/entrepreneurship-center/thought-leadership/babson-college-entrepreneurship-research-conference-bcerc/
United States Association for Small Business and Entrepreneurship (USASBE) Annual Conference	https://www.usasbe.org/page/annual-conference
American Society for Engineering Education (ASEE) Annual Conference – Entrepreneurship (ENT) Division	https://www.asee.org/events/Conferences-and-Meetings

12.3 Journals

Academic journals play a crucial role in faculty development by providing educators with research-based insights, pedagogical advancements, and evidence-backed strategies for integrating the entrepreneurial mindset into the engineering classroom. These peer-reviewed publications serve as a repository of knowledge, offering rigorous studies, case analyses, and innovative teaching methodologies that help faculty refine their instructional approaches. Moreover, journal articles present cutting-edge research on how entrepreneurial thinking can be embedded in engineering education, fostering skills such as creativity, problem-solving, and adaptability. By engaging with academic literature, faculty members can stay informed about emerging trends, evaluate the effectiveness of different instructional techniques, and contribute to the ongoing scholarly conversation. **Access to high-quality research empowers educators to make data-driven decisions**, ensuring their teaching methods align with best practices and effectively prepare students for the complexities of modern engineering careers (Table 12.2).

12.4 Professional Networks

Professional networks are essential for faculty development, as they provide educators with the connections, support, and collaborative opportunities needed to successfully integrate the entrepreneurial mindset into the engineering classroom. These networks offer platforms for faculty to exchange ideas, share resources, and learn from peers who are also working to instill entrepreneurial thinking in engineering education. Through mentorship, discussion forums, and joint research initiatives, professional networks create a community of practice where educators can gain new perspectives, refine their teaching methods, and stay informed about emerging trends. **Engaging with these networks also provides access to funding opportunities, curriculum development support, and industry collaborations that enhance the learning experience for students.** By fostering connections and knowledge-sharing, professional networks empower faculty to continuously improve their teaching strategies, ensuring that students develop the creativity, resilience, and problem-solving skills needed for success in the evolving engineering landscape (Table 12.3).

Table 12.2 Example journals

Journal	URL
Entrepreneurship Education & Pedagogy	https://us.sagepub.com/en-us/nam/entrepreneurship-education-and-pedagogy/journal203418
The Journal of Engineering Entrepreneurship	https://www.sciencegate.app/source/225747
Entrepreneurship Education	https://link.springer.com/journal/41959

Table 12.3 Example professional networks

Professional Network	URL
KEEN: The Kern Entrepreneurial Engineering Network	https://engineeringunleashed.com/what-is-keen
National Academy of Engineering Grand Challenges Scholars Program	https://viterbiundergrad.usc.edu/gcsp/
VentureWell	https://venturewell.org/
Engineering for One Planet	https://engineeringforoneplanet.org/
Biomimicry Institute	https://biomimicry.org/

Table 12.4 Example online platforms

Online platform	URL
Engineering Unleashed	https://engineeringunleashed.com/
MIT OpenCourseWare	https://ocw.mit.edu/
Stanford d.school	https://dschool.stanford.edu/
SoTL Accelerator	https://www.sotlaccelerator.com/

12.5 Online Platforms

Online platforms play a crucial role in faculty development by offering flexible, accessible, and up-to-date resources for integrating the entrepreneurial mindset into the engineering classroom. These digital hubs provide educators with instructional materials, case studies, lesson plans, and interactive tools that facilitate active learning and real-world problem-solving. These platforms **enable faculty to stay current with emerging trends**, connect with a global community of educators, and access professional development courses that enhance their teaching effectiveness. Additionally, online forums, webinars, and collaborative spaces foster knowledge-sharing and innovation, allowing instructors to refine their approaches and experiment with new pedagogical strategies. By leveraging these digital resources, educators can continually evolve their teaching practices, ensuring students develop the creativity, adaptability, and entrepreneurial mindset necessary for success in engineering and beyond (Table 12.4).

12.6 Summary

In summary, **integrating the entrepreneurial mindset into the engineering classroom requires continuous faculty development**, and a wealth of resources are available to support this effort. Conferences provide valuable opportunities for networking, collaboration, and hands-on learning, whereas academic journals offer research-based insights and evidence-backed teaching strategies. Professional networks create communities of practice where educators can exchange ideas, access mentorship, and engage in joint initiatives. Online platforms further enhance faculty development by offering flexible, on-demand resources, interactive tools, and global connections. By leveraging these diverse resources, educators can stay informed about emerging trends, refine their teaching methods, and effectively equip students with the creativity, problem-solving skills, and adaptability needed for success in the evolving engineering landscape.

MIX
Papier aus verantwortungsvollen Quellen
Paper from responsible sources
FSC® C105338

If you have any concerns about our products,
you can contact us on
ProductSafety@springernature.com

In case Publisher is established outside the EU,
the EU authorized representative is:
Springer Nature Customer Service Center GmbH
Europaplatz 3, 69115 Heidelberg, Germany

Printed by Libri Plureos GmbH
in Hamburg, Germany